made INCREDIBLY EASY!

Anatomy & Physiology

Adapted for the UK by

William N. Scott, BSc, MPhil, PhD

Lecturer in Biomedicine
Faculty of Medicine, Health
and Life Sciences
Queen's University, Belfast

First UK Edition

Wolters Kluwer | Lippincott Williams & Wilkins
Health
Philadelphia • Baltimore • New York • London
Buenos Aires • Hong Kong • Sydney • Tokyo

Staff

Acquisitions Editor
Rachel Hendrick

Academic Marketing Executive
Alison Major

Production Editor
Kevin Johnson

Proofreader
Audra Obrien

Illustrator
Bot Roda

Text and Cover Design
Designers Collective

Printed in China

For information, write to Lippincott Williams & Wilkins, 250 Waterloo Road, London SE1 8RD

British Library Cataloging-in-Publication Data. A catalogue record for this book is available from the British Library
ISBN-13: 978-1-901831-22-1
ISBN-10: 1-901831-22-1

CCS0711

Contents

Acknowledgements

Pete Bridge, BSc, MSc
Senior Lecturer, Faculty of Health and Wellbeing,
Sheffield Hallam University

Christine Caldwell, BSc, RN, MSc, PGCertHE
Senior Lecturer, Faculty of Health and Social Care,
London South Bank University

Jill Fillingham, RGN, BSc (Hons), MA and Cert Ed.
Senior Lecturer, Solstice fellow, Faculty of Health,
Edge Hill University

Declan O'Reilly, BSc (Hons), MA, PGCE, RGN
Senior Lecturer, Department of Health, Psychology &
Social Care Manchester, Metropolitan University

Katherine M.A. Rogers, BSc (Hons), PhD, PGCHET
Lecturer in Applied Health Sciences, School of Nursing and
Midwifery, The Queen's University of Belfast

David Sturgeon, BA (Hons), MPhil, DipHE, PGCLT (HE)
Senior Lecturer, Nursing and Applied Clinical Studies,
Canterbury Christ Church University

David James Tait, BSc (Hons), MSc, RMN, RGN, RNT
Lecturer, School of Nursing, Midwifery and Social Care,
Edinburgh Napier University

Foreword to the UK edition

Anatomy and physiology are an essential component of many health-related educational programmes. These subjects form a foundation for the understanding of body structure and function in health and re-enforce the importance of homeostatic imbalance in illness and disease.

Anatomy & Physiology Made Incredibly Easy, 1st UK Edition, is an excellent resource designed to help you understand the important basics of these biomedical sciences. The text serves to enhance learning and understanding through uncomplicated explanations of essential concepts and light-hearted approach to descriptions of the major body systems.

As with other texts in the successful *Made Incredibly Easy* series, this invaluable addition uses clear concise language to make the information easy to learn and remember.

Adapted specifically for UK health professionals, the book is a perfect resource both for those with no prior exposure to anatomy and physiology and for those who desire a less formal approach to revision of key topics.

The text follows a well-defined layout, each chapter focusing on an individual organ system. Objectives at the beginning of each chapter tell you what you'll learn. The body of the chapter teaches crucial information and offers a wealth of tools designed to enhance learning. Icons interspersed throughout the text highlight important information and key concepts:

 Zoom in—provides a close look at anatomic structures

 Body shop—helps explain how body systems and structures work together

 Now I get it!—converts complex physiology into easy-to-digest explanations

 As time goes by . . . —pinpoints the effects of aging on important aspects of anatomy and physiology

Memory jogger—reinforces learning through easy-to-remember anecdotes and mnemonics.

Useful quizzes at the end of each chapter help the readers validate their learning and appendices are included which give helpful information with respect to common medical definitions. Additional multiple-choice questions are included in the 'Practice makes perfect' section to further gauge learning. A comprehensive set of study cards is also included in the appendices and serves to test knowledge of the structure of the major body organs and systems.

Access to thePoint™ provides an additional dimension to the text, complimenting each chapter with useful electronic presentations, animations, a comprehensive image bank, study cards and test resources – all designed to help you get the best out of this text!

Maintaining an interest and understanding of anatomy and physiology is difficult for many students of the biomedical sciences as the volume and depth of knowledge can, at times, appear overwhelming—especially to the uninitiated or those who struggle with unfamiliar concepts and terms. However, the information contained in *Anatomy & Physiology Made Incredibly Easy*, 1st UK Edition, is as useful to the novice as it is to the expert practitioner—the brevity and clarity of content maintaining reader interest while making learning about anatomy and physiology fun!

I hope you find this book helpful in your studies. Best of luck throughout your career!

William N. Scott, BSc, MPhil, PhD
Lecturer in Biomedicine
Faculty of Medicine, Health and Life Sciences
Queen's University, Belfast

Contributors and consultants to the US edition

Katrina D. Allen, RN, MSN, CCRN
Nursing Faculty
Faulkner State Community College
Bay Minette, Ala.

Nancy Berger, RN, BC, MSN
Instructor
Charles E. Gregory School of Nursing at
Raritan Bay Medical Center
Perth Amboy, N.J.

Regina Cameron, RN, MSN, CNN
Home Hemodialysis Coordinator
Davita Dialysis
Philadelphia

Yvette P. Conley, PhD
Assistant Professor
University of Pittsburgh

Kim Cooper, RN, MSN
Nursing Department Chair
Ivy Tech Community College
Terre Haute, Ind.

Anna Easter, APRN, BC, MSN, PhD
Advanced Practice Nurse
Central Arkansas Veterans Healthcare
System
Little Rock

Rebecca Hickey, RN, AHI, CHI
Instructor
Butler Technology & Career Development
Schools
Fairfield Township, Ohio

Ruth Howell, MEd, BSN
Director, Practical Nursing Program
TriCounty Technology Center
Bartlesville, Okla.

Pamela Moody, CRNP, MSN, PhD
Nurse Administrator—Area 3
Alabama Department of Public Health
Tuscaloosa

E. Ann Myers, MD, FACP, FACE
Physician
Golden Gate Endocrinology
San Francisco

Sherry Parmenter, RD, LD
Clinical Dietitian
Fairfield Medical Center
Lancaster, Ohio

Charles W. Reick, Jr., MS, RRT
Clinical Specialist, Respiratory Care
Greater Baltimore Medical Center
Towson, Md.

**Maria Elsa Rodriguez, RN, MSN-CNS/
Education, CMSRN**
Director of Education
Kindred Hospital
San Diego

Kendra S. Seiler, RN, MSN
Nursing Instructor
Rio Hondo College
Whittier, Calif.

Denise R. York, RNC, CNS, MS, MEd
Professor
Columbus (Ohio) State Community College

1 The human body

Just the facts

In this chapter, you'll learn:

♦ anatomical terms for direction, reference planes, body cavities and body regions to help describe the locations of various body structures

♦ the structure of cells

♦ cell reproduction and energy generation

♦ four basic tissue types and their characteristics.

Anatomical terms

Anatomical terms describe directions within the body as well as the body's reference planes, cavities and regions.

Directional terms

When navigating the body, directional terms help you determine the exact location of a structure.

Couples at odds

Generally, directional terms can be grouped in pairs of opposites:
• *Superior* and *inferior* mean above and below, respectively. For example, the heart is superior to the liver, and the stomach is inferior to the lungs.
• *Anterior* means towards the front of the body, and *posterior* means towards the back. *Ventral* is sometimes used instead of anterior, and *dorsal* is sometimes used instead of posterior.
• *Medial* means towards the body's midline and *lateral* means away from it.
• *Proximal* and *distal* mean closest and farthest, respectively, to the point of origin of a structure (or to the trunk).

Locating hidden treasure, er, I mean body structures starts with directional terms, reference planes, cavities, and regions.

• *Superficial* and *deep* mean towards or at the body surface and farthest from it.

Reference planes

Reference planes are imaginary lines used to section the body and its organs. These lines run longitudinally, horizontally and angularly. The four major body reference planes are:

 sagittal

 frontal

 transverse

 oblique. (See *Picturing body reference planes*.)

Memory jogger

To remember the meanings of **proximal** and **distal**, keep in mind that when something is in **proximity**, it's nearby. When something is **distant**, it's far away.

Body shop
Picturing body reference planes

Body reference planes are used to indicate the locations of body structures. Shown here are the sagittal, frontal and transverse planes which are all at right angles to each other. An oblique plane (not shown) passes through the body or an organ at an angle between the transverse and either the frontal or sagittal planes.

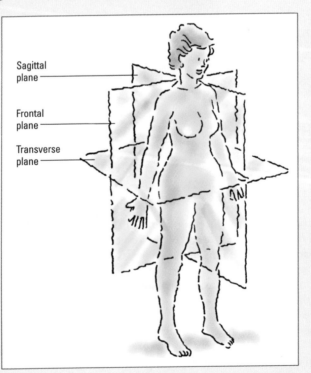

Sagittal plane

Frontal plane

Transverse plane

Body cavities

Body cavities are spaces within the body that contain the internal organs. The *dorsal* and *ventral cavities* are the two major closed cavities—cavities without direct openings to the outside of the body. (See *Locating body cavities*.)

Dorsal cavity

The dorsal cavity is located in the posterior region of the body.

The think tank and backbone of the operation

The dorsal cavity is further subdivided into the *cranial cavity*, formed by the bones of the skull and which encases the brain, and the *vertebral cavity*, formed by the vertebrae of the backbone and which encloses the spinal cord.

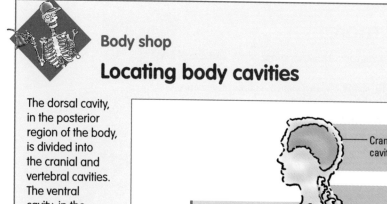

Body shop

Locating body cavities

The dorsal cavity, in the posterior region of the body, is divided into the cranial and vertebral cavities. The ventral cavity, in the anterior region, is divided into the thoracic and abdominopelvic cavities.

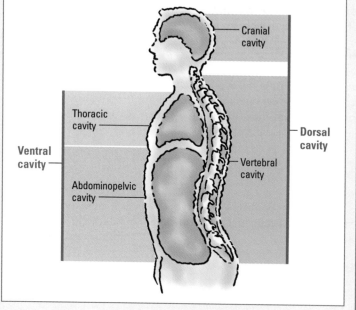

Cranial cavity

Thoracic cavity

Ventral cavity

Abdominopelvic cavity

Dorsal cavity

Vertebral cavity

Ventral cavity

The ventral cavity occupies the anterior region of the trunk. This cavity is subdivided into the *thoracic cavity* and the *abdominopelvic cavity*.

Treasure chest

> Technically speaking, the pleural cavities are only a fluid-filled space.

Surrounded by the ribs and chest muscles, the thoracic cavity refers to the space located superior to the abdominopelvic cavity. It's subdivided into the *mediastinum, pleural cavities* and *pericardial cavity:*
- The *mediastinum* houses the heart, large vessels of the heart, trachea, oesophagus, thymus, lymph nodes, and other blood vessels and nerves.
- Each of the two *pleural cavities* surrounds a lung. Each pleural cavity is a small, fluid-filled space that lies between the serous membrane that covers each lung.
- The pericardial cavity, located within the mediastinum, is also a fluid-filled space. It lies between the serous membranes that primarily cover the heart.

A six pack ... and more!

The abdominopelvic cavity has two regions, the *abdominal cavity* and the *pelvic cavity:*
- The *abdominal cavity* contains the stomach, intestines, spleen, liver and other organs.
- The *pelvic cavity*, which lies inferior to the abdominal cavity, contains the bladder, some of the reproductive organs and the rectum.

Other cavities

The body also contains an *oral cavity* (the mouth), a *nasal cavity* (located in the nose), *orbital cavities* (which house the eyes), *middle ear cavities* (which contain the small bones of the middle ear) and the *synovial cavities* (enclosed within the capsules surrounding freely moveable joints). The *sinuses* are air-filled cavities found in many of the skull and facial bones.

Body regions

Body regions are used to designate body areas that have special nerves or vascular supplies or those that perform special functions. Such areas include the cephalic (head), cervical (neck), brachial (arm) and femoral (thigh) regions, amongst others.

The guts of the matter

The most widely used body region terms are those that designate the sections of the abdomen. (See *Abdominal regions exposed*.)

The abdomen has nine regions:
- The *umbilical region*, the area around the umbilicus, includes sections of the small and large intestines, inferior vena cava and abdominal aorta.

Body shop

Abdominal regions exposed

Location of the abdominal regions, anterior aspect.

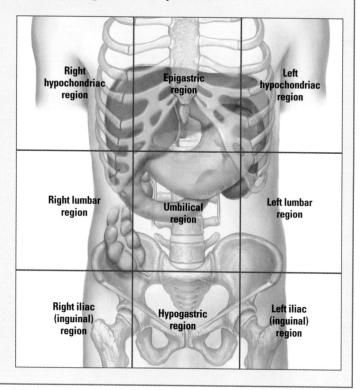

Right hypochondriac region

Epigastric region

Left hypochondriac region

Right lumbar region

Umbilical region

Left lumbar region

Right iliac (inguinal) region

Hypogastric region

Left iliac (inguinal) region

- The *epigastric region*, superior to the umbilical region, contains most of the pancreas and portions of the stomach, liver, inferior vena cava, abdominal aorta and duodenum.
- The *hypogastric region* (or pubic area), inferior to the umbilical region, houses a portion of the sigmoid colon, the urinary bladder and ureters, the uterus and ovaries (in females) and portions of the small intestine.
- The right and left *iliac regions* (or inguinal regions) are situated on either side of the hypogastric region. They include portions of the small and large intestines.
- The right and left *lumbar regions* are located on either side of the umbilical region. They include portions of the small and large intestines and portions of the kidneys.
- The right and left *hypochondriac regions*, which reside on either side of the epigastric region, contain the diaphragm, portions of the kidneys, the right side of the liver, the spleen and part of the pancreas.

Remember, each region has a specific nerve or vascular supply or performs a special function.

A look at the cell

The *cell* makes up the body's structure and serves as the basic unit of living matter. Human cells vary widely, ranging from the simple squamous epithelial cell (found in capillaries, alveoli of the lungs and the glomeruli of kidneys) to the highly specialised neurone (nerve cell).

The less complex I am, the more I can regenerate!

The greatest regeneration

Generally, the more simple the cell, the greater its power to regenerate. The more specialised the cell, the weaker its regenerative power. Cells with greater regenerative power have shorter life spans than those with less regenerative power.

Cell structure

Cells are made up of three basic components:

 cytoplasm

 plasma membrane

 nucleus. (See *Inside the cell*.)

Cytoplasm

Cytoplasm, a viscous, translucent, watery material, is the primary component of most cells. It contains a large percentage of water, inorganic ions (such as potassium, calcium, magnesium and sodium) and naturally occurring organic compounds (such as proteins, lipids and carbohydrates).

Getting charged

The inorganic ions within cytoplasm are called *electrolytes*. They regulate acid-base balance and control intracellular and extracellular water content. When these ions lose electrons (minute particles with a negative charge), they acquire a positive electrical charge. When they gain electrons, they acquire a negative electrical charge. The most common electrolytes in the body are sodium (Na^+), potassium (K^+) and chloride (Cl^-).

I'm more than just a pretty face!

A pair of 'plasms'

Nucleoplasm is similar to cytoplasm but is contained within the cell's nucleus. It plays a part in reproduction. *Cytoplasm* is the packing substance of the cell body that surrounds the nucleus and is the site of most of the cells synthesising activities. The fluid portion of cytoplasm (*cytosol*) provides

Zoom in

Inside the cell

This cross section shows the components and structures of a cell. As noted, each component plays a part in maintaining the health of the cell.

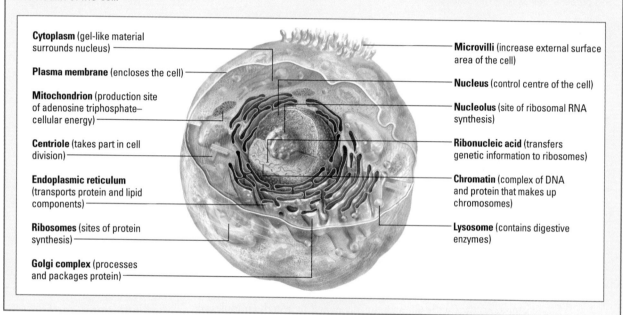

Cytoplasm (gel-like material surrounds nucleus)

Plasma membrane (encloses the cell)

Mitochondrion (production site of adenosine triphosphate–cellular energy)

Centriole (takes part in cell division)

Endoplasmic reticulum (transports protein and lipid components)

Ribosomes (sites of protein synthesis)

Golgi complex (processes and packages protein)

Microvilli (increase external surface area of the cell)

Nucleus (control centre of the cell)

Nucleolus (site of ribosomal RNA synthesis)

Ribonucleic acid (transfers genetic information to ribosomes)

Chromatin (complex of DNA and protein that makes up chromosomes)

Lysosome (contains digestive enzymes)

support for the cellular *organelles* and *inclusion bodies* that it contains.

A cytosol sea

Cytosol is a viscous, semitransparent fluid that's 70% to 90% water. It contains proteins, salts and sugars.

A lot to metabolise

• *Organelles* are the cell's metabolic units. Each organelle performs a specific function to maintain the life of the cell:
• *Mitochondria* are structures within the cytoplasm that provide most of the body's adenosine triphosphate—the enzyme that fuels many cellular activities.
• *Ribosomes* are the sites of protein synthesis.
• The *endoplasmic reticulum* is an extensive network of membrane-enclosed tubules. *Rough endoplasmic reticulum* is covered with ribosomes and produces

Now I get it!

Lysosomes at work

Lysosomes are the organelles responsible for digestion within a cell. Phagocytes assist in this process. Here's how lysosomes work.

Function of lysosomes

Lysosomes are digestive bodies that break down foreign or damaged material in cells. A membrane surrounds each lysosome and separates its digestive enzymes from the rest of the cytoplasm.

Breaking it down

The lysosomal enzymes digest matter brought into the cell by *phagocytes*, special cells that surround and engulf matter outside the cell and then transport it through the cell membrane. The membrane of the lysosome fuses with the membrane of the cytoplasmic spaces surrounding the phagocytised material; this fusion allows the lysosomal enzymes to digest the engulfed material.

certain proteins. *Smooth endoplasmic reticulum* contains enzymes that synthesise lipids.

• Each *Golgi apparatus* synthesises carbohydrate molecules. These molecules combine with the proteins produced by rough endoplasmic reticulum to form secretory products such as lipoproteins.

• *Lysosomes* are digestive bodies that break down foreign or damaged material in cells. (See *Lysosomes at work*.)

• *Peroxisomes* contain *oxidases*, enzymes capable of reducing oxygen to hydrogen peroxide and hydrogen peroxide to water.

• *Cytoskeletal elements* form a network of protein structures.

• *Centrosomes* contain *centrioles*, self-replicating organelles composed of microtubules located adjacent to the nucleus and which take part in cell division.

Temps that don't do any work

Inclusion bodies are non-functioning units in the cytoplasm that are commonly temporary. The pigment *melanin* in epidermal cells and the stored nutrient *glycogen* in liver cells are both examples of non-functioning units.

Cell membrane

The *cell membrane* is the gatekeeper of the cell. It serves as the cell's external boundary, separating it from other cells and from the external environment.

Organelles are my metabolic units.

Checkpoint

Nothing normally gets by this selectively permeable membrane without authorisation from the nucleus. The membrane consists of a double layer of phospholipids and associated protein molecules.

Nucleus

The *nucleus* is the cell's control centre. It plays a role in cell growth, metabolism and reproduction.

A nucleus may contain one or more *nucleoli*—dark-staining structures that synthesises *ribonucleic acid* (RNA). The nucleus also contains *chromosomes*. Chromosomes control cellular activity and direct protein synthesis through ribosomes in the cytoplasm.

DNA and RNA

Protein synthesis is essential for the growth of new tissue and the repair of damaged tissue. *Deoxyribonucleic acid* (DNA) carries genetic information and provides the blueprint for protein synthesis. RNA transfers this genetic information to the ribosomes, where protein synthesis occurs.

Touching all the bases

The basic structural unit of DNA is a *nucleotide*. Nucleotides consist of a phosphate group that's linked to a five-carbon sugar, *deoxyribose*, and joined to a nitrogen-containing compound called a *base*. Four different DNA bases exist:

 adenine (A)

 guanine (G)

 thymine (T)

cytosine (C)

Identifying rings

Adenine and *guanine* are double-ring compounds classified as *purines*. *Thymine* and *cytosine* are single-ring compounds classified as *pyrimidines*.

The chain gangs

DNA chains exist in pairs held together by weak chemical attractions between the nitrogen bases on adjacent chains. Because of the chemical shape of the bases, adenine bonds only with thymine and guanine bonds only with cytosine. Bases that can link with each other are called *complementary*.

Now I get it!

Types of RNA

There are three types of ribonucleic acid (RNA): ribosomal, messenger and transfer. Each has its own specific function.

Ribosomal RNA

Ribosomal RNA (rRNA) is a primary and permanent nucleic acid component of ribosomes (organelles that form the site of protein synthesis) that may be found attached to endoplasmic reticulum or free in the cytoplasm. As non-coding RNA, rRNA itself is not translated into a protein, but it does provide a mechanism for decoding messenger RNA (mRNA) into amino acids and interacting with the transfer RNAs (tRNAs) during translation.

Messenger RNA

Messenger RNA directs the arrangement of amino acids to make proteins at the ribosomes. Its single strand of nucleotides is complementary to a segment of the deoxyribonucleic acid chain that contains instructions for protein synthesis. Its chains pass from the nucleus into the cytoplasm, attaching to ribosomes there.

Transfer RNA

Transfer RNA consists of short nucleotide chains, each of which is specific for an individual amino acid. Transfer RNA carries the amino acids which match the mRNA to the ribosomes where they can be assimilated into proteins.

Insider trading

RNA consists of nucleotide chains that differ slightly from the nucleotide chains found in DNA. Several types of RNA are involved in protein synthesis. (See *Types of RNA*.)

Cell reproduction

Many cells are under a constant call to reproduce; it's either that or die. Cell division is how cells reproduce (or replicate) themselves; they achieve this through the process of *mitosis* or *meiosis*.

DNA does its thing

Before a cell divides, its genetic material is duplicated. During this process, the double helix separates into two DNA chains. Each chain serves as a template for constructing a new chain. Individual DNA nucleotides are linked into new strands with bases complementary to those in the original.

Double double

In this way, two identical double helices are formed, each containing one of the original strands and a newly formed complementary strand. These double helices are duplicates of the original DNA chain. (See *DNA up close*.)

I love creating a "new me."

Cell cycle

The *cell cycle* refers to the series of events that take place in a cell leading to its division and duplication (replication). It can be divided into two brief periods: interphase—during which the cell grows, accumulating nutrients needed for

---

Zoom in

DNA up close

Linked deoxyribonucleic acid (DNA) chains form a spiral structure, termed a *double helix*.

A twisted ladder

To understand linked DNA chains, imagine a spiral staircase. The deoxyribose and phosphate groups form the railings of the staircase, and the nitrogen base pairs (adenine and thymine, guanine and cytosine) form the steps.

Cell division

Each chain serves as a template for constructing a new chain. Before a cell divides, individual DNA nucleotides are linked into new strands with bases complementary to those in the originals. In this way, two identical double helices are formed, each containing one of the original strands and a newly formed complementary strand. These double helices are duplicates of the original DNA chain.

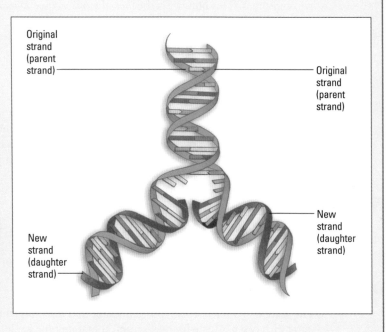

Original strand (parent strand)

Original strand (parent strand)

New strand (daughter strand)

New strand (daughter strand)

mitosis and duplicating its DNA—and the mitosis (M) phase, during which the cell splits itself into two distinct cells, often called 'daughter cells'. The cell cycle is a vital process by which a single-celled fertilised egg develops into a mature organism, as well as the process by which hair, skin, blood cells and some internal organs are renewed.

Mitosis

Mitosis is the equal division of material in the nucleus (*karyokinesis*) followed by division of the cell body (*cytokinesis*). It's the preferred mode of replication by all cells in the human body, except the gametes (sex cells). Cell division occurs in five phases, a relatively inactive preparation phase called *interphase*, and four active phases:

 prophase

 metaphase

 anaphase

 telophase

He may think he's better than me, but we have identical DNA!

Two daughters equal 46

Mitosis results in two daughter cells (exact duplicates), each containing 23 pairs of chromosomes—or 46 individual chromosomes. This number is the *diploid number*. (See *Divide and conquer: Five stages of mitosis.*)

Mitosis is a key part of the cell cycle—the coordinated duplication of a cell and its various components.

Meiosis

Meiosis is reserved for gametes (ova and spermatozoa). This process intermixes genetic material between homologous chromosomes (matched pairs of chromosomes; one from each parent), producing four daughter cells, each with the *haploid number* of chromosomes (23, or half of the 46). Meiosis has two divisions separated by a resting phase.

First division

The first division has six phases and begins with one parent cell. When the first division ends, the result is two daughter cells—each containing the haploid (23) number of chromosomes. This stage is often referred to as *reduction division*.

Division 2, the sequel

The second division is a four-phase division that resembles mitosis. It starts with two new daughter cells, each containing the haploid number of chromosomes, and ends with four new haploid cells. In each cell, the

Memory jogger

To help you remember the difference between haploid and diploid, think of the prefix **di-** in diploid. **Di-** means **double**, so diploid cells have double the number of chromosomes in a haploid cell.

Now I get it!

Divide and conquer: The five stages of mitosis

Through the process of mitosis, the nuclear content of all body cells (except gametes) reproduces and divides. The result is the formation of two new daughter cells, each containing the diploid (46) number of chromosomes.

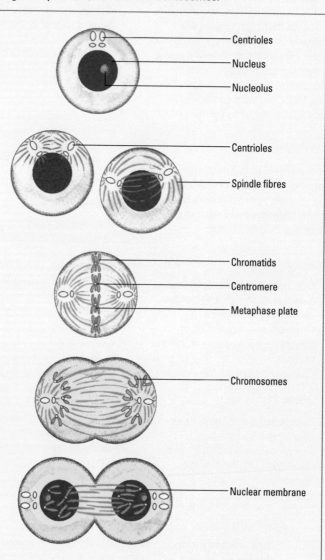

Interphase (preparation stage)

During *interphase,* the nucleus and nuclear membrane are well defined, and the nucleolus is visible. As chromosomes replicate, each forms a double strand that remains attached at the centre by a centromere.

Prophase

In *prophase,* the nucleolus disappears and the chromosomes become distinct. *Chromatids,* halves of each duplicated chromosome, remain attached by a centromere. Centrioles move to opposite sides of the cell and radiate spindle fibres.

Metaphase

Metaphase occurs when chromosomes line up randomly in the centre of the cell between the spindles, along the *metaphase plate.*

Anaphase

Anaphase is characterised by centromeres moving apart, pulling the separate chromatids (now called *chromosomes*) to opposite ends of the cell. The number of chromosomes at each end of the cell equals the original number.

Telophase

During *telophase,* the final stage of mitosis, a nuclear membrane forms around each nucleus and spindle fibres disappear. The cytoplasm compresses and divides the cell in half. Each new cell contains the diploid (46) number of chromosomes.

Now I get it!

Meiosis: Step-by-step

Meiosis has two divisions that are separated by a resting phase. By the end of the first division, two daughter cells exist that each contain the haploid (23) number of chromosomes. When the second division ends, each of the two daughter cells from the first division divides, resulting in four daughter cells, each containing the haploid number of chromosomes.

First division

Preparatory events (interphase) precede the first division, which has four phases. Here's what happens during interphase and each of the phases in the first division.

Interphase

1. Chromosomes replicate, forming a double strand attached at the centre by a centromere.
2. Chromosomes appear as an indistinguishable matrix within the nucleus.
3. Centrioles appear outside the nucleus.

Prophase I

1. The nucleolus and the nuclear membrane disappear.
2. Chromosomes are distinct, with chromatids attached by the centromere.

3. Homologous chromosomes move close together and intertwine; exchange of genetic information (genetic recombination) may occur.
4. Centrioles separate and spindle fibres appear.

Metaphase I

1. Pairs of homologous chromosomes line up randomly along the metaphase plate.
2. Spindle fibres attach to each chromosome pair.

Anaphase I

1. Homologous pairs separate.
2. Spindle fibres pull homologous, double-stranded chromosomes to opposite ends of the cell.
3. Chromatids remain attached.

Telophase I

1. The nuclear membrane forms.
2. Spindle fibres and chromosomes disappear.
3. Cytoplasm compresses and divides the cell in half.
4. Each new cell contains the haploid (23) number of chromosomes.

A short preparation phase (interkinesis) precedes the second division.

Second division

The second division closely resembles mitosis and is characterised by these four phases.

Prophase II

1. The nuclear membrane disappears.
2. Spindle fibres form.
3. Double-stranded chromosomes appear as thin threads.

Metaphase II

1. Chromosomes line up along the metaphase plate.
2. Centromeres replicate.

Anaphase II

1. Chromatids separate (now a single-stranded chromosome).
2. Chromosomes move away from each other to the opposite ends of the cell.

Telophase II

1. The nuclear membrane forms.
2. Chromosomes and spindle fibres disappear.
3. Cytoplasm compresses, dividing the cell in half.
4. Four daughter cells are created, each of which contains the haploid (23) number of chromosomes.

two chromatids of each chromosome separate to form new daughter cells. However, because each cell entering the second division has only 23 chromosomes, each daughter cell formed has only 23 chromosomes. (See *Meiosis: step-by-step*.)

Cellular energy generation

All cellular function depends on energy generation and transportation of substances within and among cells.

Cellular power

Adenosine triphosphate (ATP) serves as the chemical fuel for cellular processes. ATP consists of a nitrogen-containing compound (adenine) joined to a five-carbon sugar (ribose), forming adenosine. Adenosine is joined to three phosphate (or triphosphate) groups. Chemical bonds between the first and second phosphate groups and between the second and third phosphate groups contain abundant energy.

The three Rs

ATP needs to be converted to *adenosine diphosphate* (ADP) to produce energy. To understand this conversion, remember the three Rs:
• *Rupture*—ATP is converted to ADP when the terminal high-energy phosphate bond ruptures.
• *Release*—Because the third phosphate is liberated, energy stored in the chemical bond is released.
• *Recycle*—Mitochondrial enzymes then reconvert ADP and the liberated phosphate to ATP. To obtain the energy needed for this reattachment, mitochondria oxidise nutrients. This makes recycled ATP available again for energy production.

Movement within cells

Each cell interacts with body fluids through the interchange of substances.

Modes of transportation

Several transport methods—*diffusion*, *osmosis* and *active transport processes*—move substances between cells and body fluids. In another method, *filtration* (which occurs in the glomerular capsule of the kidney nephrons), fluids and dissolved substances are transferred across capillaries into *interstitial fluid* (fluid in the spaces between cells and tissues).

Diffusion
In *diffusion*, solutes move from an area of higher concentration to one of lower concentration. Eventually, an equal distribution of solutes between the two areas occurs.

Go with the flow

Diffusion is a form of passive transport—no energy is required to make it happen; movement of solutes occurs from

No energy is required for diffusion, so I just go with the flow.

Now I get it!

Understanding passive transport

No energy is required for passive transport. It occurs through two mechanisms: diffusion and osmosis.

Diffusion

In diffusion, substances move from an area of higher concentration to an area of lower concentration. Movement continues until distribution is uniform.

Osmosis

In osmosis, fluid moves from an area of higher concentration to one of lower concentration.

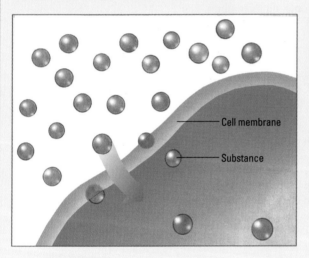

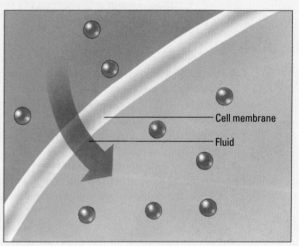

an area of high concentration to an area of low concentration until a balance occurs. (See *Understanding passive transport*.)

Advancing and declining rates

Several factors influence the rate of diffusion:
* *concentration gradient*—the greater the concentration gradient (the difference in particle concentration on either side of the plasma membrane), the faster the rate of diffusion
* *particle size*—the smaller the particles, the faster the rate of diffusion
* *lipid solubility*—the more lipid soluble the particles are, the more rapidly they diffuse through the lipid layers of the cell membrane
* *membrane permeability*—the more permeable the membrane, the greater the rate of diffusion.

Osmosis

Osmosis is the passive transport of fluid across a membrane, from an area of lower solute concentration (comparatively *more* fluid) into an area of higher solute concentration (comparatively *less* fluid).

Enough is enough

Osmosis stops when enough fluid has moved through the membrane to equalise the solute concentration on both sides of the membrane.

Active transport

Active transport requires energy. Usually, this mechanism moves a substance across the cell membrane against the concentration gradient—from an area of lower concentration to one of higher concentration.

ATP at it again

The energy required for a solute to move against a concentration gradient comes from ATP. ATP is stored in all cells and supplies energy for solute movement in and out of cells. (See *Understanding active transport*, page 18.)

It goes both ways

However, active transport also can move a substance with the concentration gradient. In this process, a carrier molecule in the cell membrane combines with the substance and transports it through the membrane, depositing it on the other side.

Endocytosis

Endocytosis is an active transport method in which, instead of passing through the cell membrane, a substance is engulfed by the cell. The cell surrounds the substance with part of the cell membrane. This part separates to form a *vacuole* (cavity) that moves to the cell's interior.

Gobbling up particles

Endocytosis involves either *phagocytosis* or *pinocytosis*. Phagocytosis refers to engulfment and ingestion of particles that are too large to pass through the cell membrane. Pinocytosis occurs only to engulf dissolved substances or small particles suspended in fluid.

Filtration

Fluid and dissolved substances also may move across a cell membrane by *filtration*.

Several factors influence the rate of diffusion—concentration gradient, particle size, lipid solubility and membrane permeability.

Active transport requires energy. As a cell, that means I need to fill up on ATP.

Now I get it!

Understanding active transport

Active transport moves molecules and ions against a concentration gradient from an area of lower concentration to one of higher concentration. This movement requires energy, usually in the form of adenosine triphosphate (ATP). The sodium-potassium pump and pinocytosis are examples of active transport mechanisms.

Sodium-potassium pump

The sodium-potassium pump moves sodium from inside the cell to outside, where the sodium concentration is greater; potassium moves from outside the cell to inside, where the potassium concentration is greater.

Pinocytosis

In pinocytosis, tiny vacuoles take droplets of fluid containing dissolved substances into the cell. The engulfed fluid is used in the cell.

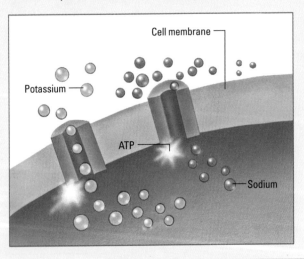

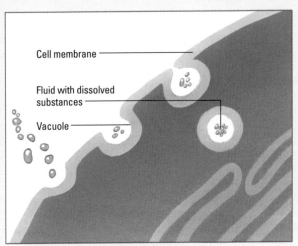

Pressure is the point

In filtration, pressure (provided by capillary blood) is applied to a solution on one side of the cell membrane. The pressure forces fluid and dissolved particles through the membrane. The rate of filtration (how quickly substances pass through the membrane) depends on the amount of pressure. Filtration promotes the transfer of fluids and dissolved materials from the blood across the capillaries into the interstitial fluid. This process occurs within the glomerular apparatus of the kidney nephrons.

A look at human tissue

Tissues are groups of cells that perform the same general function. The human body contains four basic types: *epithelial*, *connective*, *muscle* and *nervous tissue*.

A group of cells working together on the same function is called a tissue.

Epithelial tissue

Epithelial tissue (epithelium) is a continuous cellular sheet that covers the body's surface, lines body cavities and forms certain glands. (See *Distinguishing types of epithelial tissue*, page 20.)

Patrolling the borders

In the tubules of the kidneys and the internal mucosa of small intestine, borders of columnar epithelial cells have tiny, brushlike structures (microvilli) called a *brush border*.

This hair isn't just for looks

Some epithelial cells possess *cilia*, fine protuberances that are larger than microvilli and move fluid and particles through the cavity of an organ.

Endothelium

Epithelial tissue with a single layer of squamous cells attached to a basement membrane is called *endothelium*. Such tissue lines the heart, blood and lymphatic vessels, and other organs.

Glandular epithelium

Organs that produce secretions consist of a special type of epithelium called *glandular epithelium*.

The secret is in how it secretes

Glands are classified as endocrine or exocrine according to how they secrete their products.
- *Endocrine glands* release their secretions into extracellular fluids. For instance, the medulla of the adrenal gland secretes adrenaline (epinephrine) and noradrenaline (norepinephrine) into the bloodstream.
- *Exocrine glands* discharge their secretions into ducts that lead to external or internal surfaces. For example, the sweat glands secrete sweat onto the surface of the skin.

Zoom in

Distinguishing types of epithelial tissue

Epithelial tissue (epithelium) is classified by the number of cell layers and the shape of surface cells. Some types of epithelium go through a process of desquamation (shedding of surface cells) and regenerate continuously by transformation of cells from deeper layers.

Identified by number of cell layers

Classified by the number of cell layers, epithelium may be *simple* (one-layered), *stratified* (multilayered) or *pseudostratified* (one-layered but appearing to be multilayered).

Classified by shape

If classified by shape, epithelium may be *squamous* (containing flat surface cells), *columnar* (containing tall, cylindrical surface cells) or *cuboidal* (containing cube-shaped surface cells).

The top left illustration below shows how the basement membrane of simple squamous epithelium joins the epithelium to underlying connective tissues. The remaining illustrations show the five other types of epithelial tissue.

Simple squamous epithelium
Single layer of flattened cells with disc-shaped nuclei; found in capillaries, alveoli and glomeruli, and adapted for diffusion or absorption.

Simple columnar epithelium
Single layer of tall cells with oval nuclei; found typically lining the stomach and intestines and involved in absorption, secretion and protection.

Stratified squamous epithelium
Basal cells that are cuboidal or columnar; found as multilayered component of skin, or lining of the oesophagus where they protect from abrasion.

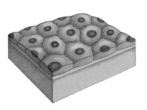
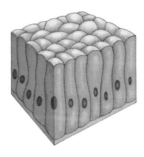
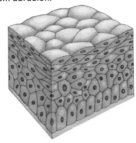

Simple cuboidal epithelium
Single layer of cube-like cells; lines ducts of excretory glands and kidney tubules where they have secretory or absorptive roles.

Stratified columnar epithelium
Superficial cells that are elongated and columnar; found primarily in the conjunctiva, pharynx and anus where they function in aspects of protection and secretion.

Pseudostratified columnar epithelium
Cells of different height with nuclei at different levels; typically found in the lining of the trachea and involved in secretion or absorption.

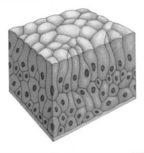
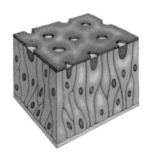
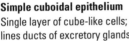

Mixing it up

Mixed glands contain both endocrine and exocrine cells. The pancreas is a mixed gland. As an endocrine gland, it produces insulin, glucagon and other hormones. As an exocrine gland, it introduces pancreatic juices into the intestines.

Connective tissue

Connective tissue—a category that includes bone, cartilage, blood and adipose (fatty) tissue—binds together and supports body structures. Connective tissue is classified as *loose* or *dense*.

> In addition to bones, connective tissue includes cartilage, blood and adipose tissue.

Cut loose

Loose (areolar) connective tissue has large spaces that separate the fibres and cells. It contains a lot of intercellular fluid.

Dense tissue issues

Dense connective tissue provides structural support and has greater fibre concentration. Dense tissue is further subdivided into dense regular and dense irregular connective tissue:
*   *Dense regular* connective tissue consists of tightly packed fibres arranged in a consistent pattern. It includes tendons, ligaments and aponeuroses (flat fibrous sheets that attach muscles to bones or other tissues).
*   *Dense irregular* connective tissue has tightly packed fibres arranged in an inconsistent pattern. It's found in the dermis, submucosa of the GI tract, fibrous capsules and fasciae.

Suppose it's adipose

Commonly called *fat, adipose tissue* is a specialised type of loose connective tissue where a single lipid (fat) droplet occupies most of each cell. Widely distributed subcutaneously, it acts as insulation to conserve body heat, as a cushion for internal organs and as a storage depot for excess food and reserve supplies of energy.

Muscle tissue

Muscle tissue consists of muscle cells with a generous blood supply. Some muscle cells can measure up to several centimetres long and have an elongated shape that enhances their *contractility* (ability to contract).

> What happened to your adipose tissue?

> I went overboard on the dieting thing.

The tissues at issue

There are three basic types of muscle tissue:

Striated (skeletal) muscle tissue gets its name from its striped, or striated, appearance; it contracts voluntarily. The cells that comprise this type of tissue are frequently multinucleated.

Cardiac muscle tissue is generally composed of cells that are branched with a striped appearance. The cells contain a single nucleus, and contract involuntarily.

Smooth-muscle tissue consists of long, spindle-shaped cells and lacks the striped pattern of skeletal and cardiac muscle tissue. Its activity is stimulated by the autonomic nervous system and isn't under voluntary control.

Wall tissue paper

Smooth-muscle tissue is found in the walls of the epithelial tissues that line the internal organs and other structures, including the respiratory passages from the trachea to the alveolar ducts, the urinary and genital ducts, the arteries and veins, the larger lymphatic trunks, the intestines, the arrectores pilorum and the iris and ciliary body of the eye.

Nervous tissue

The main function of *nervous tissue* is communication. Its primary properties are *irritability* (the capacity to react to various physical and chemical agents) and *conductivity* (the ability to transmit the resulting reaction from one point to another).

Nervous tissue specialists

Neurones are highly specialised cells that generate and conduct nerve impulses. A typical neurone consists of a cell body with cytoplasmic extensions—numerous *dendrites* on one pole and a single *axon* on the opposite pole. These extensions allow the neurone to conduct impulses over long distances.

Protecting neurones

Neuroglia form the support structure of nervous tissue, insulating and protecting neurones. They're found only in the central nervous system.

Irritable? I'll show you irritable!

Quick quiz

1.   The reference plane that divides the body lengthwise into right and left regions is the:
   A.   frontal plane.
   B.   sagittal plane.
   C.   transverse plane.
   D.   oblique plane.

Answer: B. Imaginary lines called *reference planes* are used to section the body. The sagittal plane runs lengthwise and divides the body into right and left regions.

2.   The structure that plays the biggest role in cellular function is the:
   A.   nucleus.
   B.   Golgi apparatus.
   C.   ribosome.
   D.   mitochondrion.

Answer: A. Serving as the cell's control centre, the nucleus plays a role in cell growth, metabolism and reproduction.

3.   The four basic types of tissue that the human body contains are:
   A.   muscle, cartilage, glandular and connective tissue.
   B.   bone, cartilage, glands and adipose tissue.
   C.   loose, dense connective, dense regular and dense irregular tissue.
   D.   epithelial, connective, muscle and nervous tissue.

Answer: D. Tissues are groups of cells with the same general function. The human body contains four basic types of tissue: epithelial, connective, muscle and nervous tissue.

4.   Meiosis ends when:
   A.   two new daughter cells form, each with the haploid number of chromosomes.
   B.   one daughter cell forms and is an exact copy of the original.
   C.   four new daughter cells form, each with the haploid number of chromosomes.
   D.   four new daughter cells form, each with the diploid number of chromosomes.

Answer: C. Meiosis comes to completion with the end of telophase II. The result is four daughter cells, each of which contains the haploid (23) number of chromosomes.

Scoring

☆☆☆   If you answered all four questions correctly, fantastic! You're well on your way to a fantastic voyage through the human body.

☆☆   If you answered three questions correctly, all right! You're in for some smooth sailing. Pretty soon you'll know the body inside and out.

☆   If you answered fewer than three questions correctly, it's time to structure your revision. With plenty more Quick quizzes to go, you'll conquer this body of knowledge in no time.

Just the facts

In this chapter, you'll learn:

♦ the chemical composition of the body

♦ the structure of an atom

♦ differences between inorganic and organic compounds.

A look at body chemistry

The human body is composed of chemicals; in fact, all of its activities are chemical in nature. To understand the human body and its functions, you must have an appreciation and understanding of chemistry.

Comes down to chemistry

The chemical level is the simplest and most important level of structural organisation. Without the proper chemicals in the proper amounts, body cells—and eventually the body itself—would die.

Principles of chemistry

Every cell contains thousands of different chemicals that constantly interact with one another. Differences in chemical composition differentiate types of body tissue. Furthermore, the blueprints of heredity (deoxyribonucleic acid [DNA] and ribonucleic acid [RNA]) are encoded in chemical form.

What's the matter?

Matter is anything that has mass and occupies space. It may be a solid, liquid or gas.

Without chemicals in the proper amounts, I would die.

Energetic types

Energy is the capacity to do work—to put mass into motion. It may be *potential energy* (stored energy) or *kinetic energy* (the energy of motion). Types of energy include chemical, electrical and radiant.

Chemical composition

An *element* is matter that can't be broken down into simpler substances by normal chemical reactions. All forms of matter are composed of chemical elements. Each of the chemical elements in the periodic table has a chemical symbol. For example, N is the chemical symbol for nitrogen. (See *Understanding elements and compounds.*)

It's elementary

Carbon, hydrogen, nitrogen and oxygen account for 96% of the body's total weight. Calcium and phosphorus account for another 2.5%. (See *What's a body made of?* page 28.)

Atomic structure

An *atom* is the smallest unit of matter that can take part in a chemical reaction. Atoms of a single type constitute an element.

Subatomic particles

Each atom has a dense central core called a *nucleus*, plus one or more surrounding energy layers called *electron shells*. Atoms consist of three basic subatomic particles: *protons*, *neutrons* and *electrons*. The exception to this is the atom hydrogen, which only contains a proton and an electron.

Weighing in

A proton weighs nearly the same as a neutron, and a proton and a neutron each weigh 1,836 times as much as an electron.

Protons

Protons are closely packed particles in the atom's nucleus that have a positive charge. Each element has a distinct number of protons.

Now I get it!

Understanding elements and compounds

It can be confusing to understand the difference between elements and compounds. The best way to remember the difference is to understand how atoms combine to form each.

Get at them atoms

A single atom constitutes an element. Thus, an atom of hydrogen is the element hydrogen. Now, here's the confusing part: an element can also be composed of more than one atom—a molecule.

Yep. I'm an atom—the smallest unit of matter that can take part in a chemical reaction.

Molecules: two (or more) of the same

A molecule is a combination of two or more atoms. If these atoms are the same—that is, all the same element (such as all hydrogen atoms)—they're considered a *molecule* of that element (a molecule of hydrogen).

We're atoms that are joined together to make a molecule. We're the same, so we're a molecule of an element.

Compounds: Where the different come together

If the atoms combined are different—that is, different elements (such as a carbon atom and an oxygen atom)—the molecule formed is a *compound* (such as the compound CO, or carbon monoxide).

I know we just met, but I think we'll make a beautiful compound together.

Positive thinking

An element's number of protons determines its *atomic number* and confers a positive charge. For example, all carbon atoms—and *only* carbon atoms—have six protons; therefore, the atomic number of carbon is 6.

Body Shop

What's a body made of?

This chart shows the chemical elements of the human body in descending order from most to least plentiful.

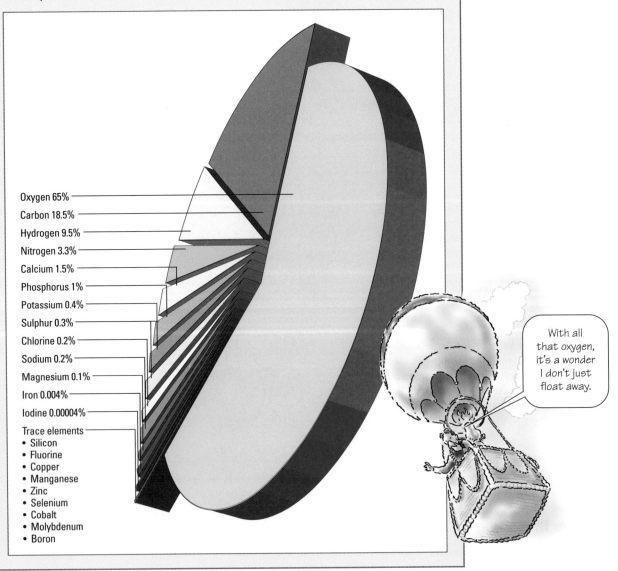

Oxygen 65%
Carbon 18.5%
Hydrogen 9.5%
Nitrogen 3.3%
Calcium 1.5%
Phosphorus 1%
Potassium 0.4%
Sulphur 0.3%
Chlorine 0.2%
Sodium 0.2%
Magnesium 0.1%
Iron 0.004%
Iodine 0.00004%
Trace elements
• Silicon
• Fluorine
• Copper
• Manganese
• Zinc
• Selenium
• Cobalt
• Molybdenum
• Boron

With all that oxygen, it's a wonder I don't just float away.

Neutrons

Neutrons are uncharged, or neutral, particles in the atom's nucleus.

Mass numbers

An atom's *mass number* is distinct from its atomic number. The atomic mass number is the sum of the number of protons and neutrons in the nucleus of an atom. You can also think of the mass number as the sum of the masses of protons and neutrons. For example, helium, with two protons and two neutrons, has a mass number of 4.

Isolating the isotopes

Not all the atoms of an element necessarily have the same number of neutrons. An *isotope* is a form of an atom that has a different number of neutrons and, therefore, a different atomic weight. Some isotopes are radioactive and this property is exploited in both the diagnosis and treatment of many diseases.

A weighty matter

Understanding isotopes is a key to another important concept, *atomic weight*. An atom's atomic weight is the average of the relative weights (mass numbers) of all the element's isotopes. (Recall that isotopes are different atomic forms of the same element that vary in the number of neutrons they contain.)

Electrons

Electrons are negatively charged particles that orbit the nucleus in electron shells. They play a key role in chemical bonds and reactions.

Staying neutral

The number of electrons in an atom equals the number of protons in its nucleus. The electrons' negative charges cancel out the protons' positive charges, making atoms electrically neutral.

Shell games

Electrons circle the nucleus in *shells* often depicted as concentric circles. Each electron shell can hold a maximum number of electrons and represents a specific energy level. The innermost shell can accommodate two electrons at most, whereas the outermost shells can hold many more.

An atom with single (unpaired) electrons orbiting in its outermost electron shell can be chemically *active*—that is, able to take part in chemical reactions. An atom with an outer shell that contains only pairs of electrons is chemically inactive, or *stable*.

The value of valency

The valency of an atom (its ability to combine with other atoms) equals the number of unpaired electrons in its outer shell. For example, sodium (Na^+) has a plus-one valence because its outer shell contains an unpaired electron.

Chemical bonds

A *chemical bond* is a force of attraction that binds a molecule's atoms together. Formation of a chemical bond usually requires energy. Breakup of a chemical bond usually releases energy.

The name's bond ...

Several types of chemical bonds exist:
- A *hydrogen bond* occurs when two atoms associate with a hydrogen atom. Oxygen and nitrogen, for instance, commonly form hydrogen bonds.
- An *ionic* (electrovalent) *bond* occurs when valence electrons transfer from one atom to another.
- A *covalent bond* forms when atoms share pairs of valence electrons. (See *Picturing ionic and covalent bonds.*)

Chemical reactions

A *chemical reaction* involves unpaired electrons in the outer shells of atoms. In this reaction, one of two events occurs:

☝ Unpaired electrons from the outer shell of one atom transfer to the outer shell of another atom.

✌ One atom shares its unpaired electrons with another atom.

How will they react?

Energy, particle concentration, speed, and orientation determine whether a chemical reaction will occur. The four basic types of chemical reactions are *synthesis, decomposition, exchange* and *reversible reactions*. (See *Comparing chemical reactions*, page 32.)

It takes energy to bind atoms together. Breaking the bond releases energy.

Now I get it!

Picturing ionic and covalent bonds

A chemical bond is a force of attraction that binds the atoms of a molecule together. Let's examine ionic and covalent bonds.

Ionic bonds

In an ionic bond, an electron is transferred from one atom to another. By forces of attraction, an electron is transferred from a sodium (Na) atom to a chlorine (Cl) atom. The result is a molecule of sodium chloride (NaCl).

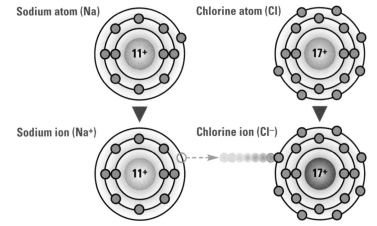

Sodium atom (Na)   Chlorine atom (Cl)

Sodium ion (Na$^+$)   Chlorine ion (Cl$^-$)

Covalent bonds

In a covalent bond, atoms share a pair of electrons. This is what happens when two hydrogen (H) atoms form a covalent bond.

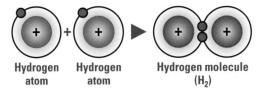

Hydrogen atom + Hydrogen atom ▶ Hydrogen molecule (H$_2$)

Now I get it!

Comparing chemical reactions

When chemical reactions occur, they involve unpaired electrons in the outer shells of atoms. Here are the four basic types of chemical reactions.

Synthesis reaction (anabolism)

A synthesis reaction combines two or more substances (reactants) to form a new, more complex substance (product). This results in a chemical bond.

$$A + B \rightarrow A B$$

Decomposition reaction (catabolism)

In a decomposition reaction, a substance decomposes, or breaks down, into two or more simpler substances, leading to the breakdown of a chemical bond.

$$A B \rightarrow A + B$$

Exchange reaction

An exchange reaction is a combination of a decomposition and a synthesis reaction. This reaction occurs when two complex substances decompose into simpler substances. The simple substances then join (through synthesis) with different simple substances to form new complex substances.

$$A B + C D \rightarrow A + B + C + D \rightarrow A D + B C$$

Reversible reaction

In a reversible reaction, the product reverts to its original reactants, and vice versa. Reversible reactions may require special conditions, such as heat or light.

$$A + B \leftrightarrow A B$$

Inorganic and organic compounds

Although most biomolecules (molecules produced by living cells) form *organic compounds*, or compounds containing carbon, some form *inorganic compounds*, or compounds without carbon.

Inorganic compounds

Inorganic compounds are usually small and include water and *electrolytes*—inorganic acids, bases and salts.

The body's reservoir

Water is the body's most abundant substance. It performs a host of vital functions, including:
• easily forming polar covalent bonds (which permits the transport of solvents)
• acting as a lubricant in mucus and other bodily fluids
• entering into chemical reactions, such as nutrient breakdown during digestion
• enabling the body to maintain a relatively constant temperature (by both absorbing and releasing heat slowly).

I need water to stay at a constant temperature.

Recognising ionising

Acids, bases and salts are *electrolytes*—compounds whose molecules consist of positively charged ions (*cations*) and negatively charged ions (*anions*) that *dissociate* (separate into ions) in solution:
• *Acids* ionise into hydrogen ions (H^+) and anions. In other words, acids separate into a positively charged hydrogen ion and a negatively charged anion.
• *Bases*, in contrast, ionise into hydroxide ions and cations. Bases separate into negatively charged hydroxide ions (OH^-) and positively charged cations.
• *Salts* form when acids react with bases. In water, salts ionise into cations and anions, but not hydrogen or hydroxide ions.

A balancing act

Body fluids must attain acid-base balance to maintain *homeostasis* (the dynamic equilibrium of the body). A solution's acidity is determined by the number of hydrogen ions it contains. The more hydrogen ions present, the more acidic the solution. Conversely, the more hydroxide ions a solution contains, the more basic, or *alkaline*, it is. The acidity or alkalinity of a solution is indicated using the pH scale (a measure of the concentration of hydrogen ions in a solution that runs from pH 1 to 14). A pH of 7 indicates a neutral solution, a pH below 7 indicates acidity, and a pH in excess of 7 indicates alkalinity.

Organic compounds

Most biomolecules form *organic compounds*—compounds that contain carbon or carbon-hydrogen bonds (with the exception of oxides, carbonates and bicarbonates). *Carbohydrates*, *lipids*, *proteins* and *nucleic acids* are all examples of organic compounds.

I'm a carbohydrate. Count on me to release energy.

Carbohydrates

In the body, *carbohydrates* are sugars, starches and glycogen.

The energy company

The main functions of carbohydrates are to release energy and store energy. There are three types of carbohydrates:

Monosaccharides, such as glucose, fructose, galactose, ribose and deoxyribose, are sugars with three to seven carbon atoms.

Disaccharides, such as lactose and maltose, contain two monosaccharides.

Polysaccharides, such as glycogen, are large carbohydrates composed of many monosaccharides.

Lipids

Lipids are water-insoluble biomolecules commonly referred to as fats. The major lipids are *triglycerides*, *phospholipids*, *steroids*, *lipoproteins* and *eicosanoids*.

To insulate and protect

Triglycerides are the most abundant class of lipid in both food and the body. These lipids are neutral fats that insulate and protect. They also serve as the body's most concentrated energy source. Triglycerides contain three molecules of a fatty acid chemically joined to one molecule of glycerol.

Bars on the cell

Phospholipids are the major structural components of cell membranes and consist of one molecule of glycerol, two molecules of a fatty acid, and a phosphate group.

No fat in cholesterol?

Steroids are simple lipids with no fatty acids in their molecules. They fall into four main categories, each of which performs different functions.
- *Bile salts* emulsify fats during digestion and aid absorption of the fat-soluble vitamins (vitamins A, D, E and K).
- *Hormones* are chemical substances that have a specific effect on other cells.
- *Cholesterol*, a part of animal cell membranes, is needed to form all other steroids.
- *Vitamin D* helps regulate the body's calcium concentration.

Porters and other hardworking lipids

Lipoproteins help transport lipids to various parts of the body. *Eicosanoids* include *prostaglandins*, which, among other functions, modify hormone responses, promote the inflammatory response and open the airways, and *leukotrienes*, which also play a part in allergic and inflammatory responses.

Memory jogger

To recall the distinction between the three types of carbohydrates, remember the prefixes:

Mono—means one.

Di—means two; therefore, disaccharides contain two monosaccharides.

Poly—means many, so you can expect that polysaccharides contain many monosaccharides.

Bile salts aid absorption of us fat-soluble vitamins.

Proteins

Proteins are the most abundant organic compound in the body. They're composed of building blocks called *amino acids*. Amino acids are linked together by *peptide bonds*—chemical bonds that join the carboxyl group of one amino acid to the amino group of another.

I'm a protein. I'm built from blocks of amino acids that form polypeptides.

Building up the blocks

Many amino acids linked together form a *polypeptide*. One or more polypeptides form a protein. The sequence of amino acids in a protein's polypeptide chain contributes to its shape. A protein's shape determines which of its many functions it performs:

- providing structure and protection
- promoting muscle contraction
- transporting various substances
- regulating processes
- forming antibodies that help protect against infection
- serving as an enzyme (the largest group of proteins, which act as catalysts for crucial chemical reactions).

Nucleic acids

The nucleic acids DNA and RNA are composed of nitrogenous bases, sugars and phosphate groups. The primary hereditary molecule, DNA, contains two long chains of deoxyribonucleotides, which coil into a double-helix shape.

Holding it together

Deoxyribose and phosphate units alternate in the 'backbone' of the chains. Holding the two chains together are base pairs of adenine-thymine and guanine-cytosine.

RNA and its special function

Unlike DNA, RNA has a single-chain structure. It contains ribose instead of deoxyribose and replaces the base thymine with uracil. RNA transmits genetic information from the cell nucleus to the cytoplasm. In the cytoplasm, it guides protein synthesis from amino acids.

Quick quiz

1. The three most plentiful chemical elements in the human body are:
    A. phosphorus, hydrogen and oxygen.
    B. carbon, oxygen and silicon.
    C. oxygen, carbon and hydrogen.
    D. oxygen, carbon and nitrogen.

Answer: C. There are 22 chemical elements in the human body. Oxygen (65%), carbon (18.5%) and hydrogen (9.5%) are the three most plentiful.

2. Protons are closely packed particles in the atom's nucleus that have:
    A. a positive charge.
    B. a negative charge.
    C. a neutral charge.
    D. a mixed charge.

Answer: A. Protons are positively charged particles in the atom's nucleus.

3. Which of the following is an example of an organic compound?
    A. water.
    B. an electrolyte.
    C. a protein.
    D. an acid.

Answer: C. Organic compounds are compounds that contain carbon. Examples include carbohydrates, lipids, proteins and nucleic acids.

Scoring

☆☆☆   If you answered all three questions correctly, congratulations. You and this chapter have achieved homeostasis.

☆☆   If you answered two questions correctly, excellent. You and this chapter go together as neatly as amino acids forming a polypeptide chain.

☆   If you answered only one question correctly, no worries. Get yourself organised and then go back and review the chapter.

Just the facts

In this chapter, you'll learn:

♦ major muscles and bones of the body

♦ types of muscle tissue and their functions

♦ types of bones and their functions

♦ the roles of tendons, ligaments, cartilage, joints, and bursae in body movement and structure

A look at the musculoskeletal system

The musculoskeletal system consists of muscles, tendons, ligaments, bones, cartilage, joints and bursae. These structures give the human body its shape and ability to move.

How the body moves

Various parts of the musculoskeletal system work with the nervous system to produce voluntary movements. Muscles contract when stimulated by impulses from the nervous system.

Using the force

During contraction, the muscle shortens, pulling on the bones to which it's attached. This causes movement of one of the bones; most movements involve groups of muscles rather than one muscle.

Structures of the musculoskeletal system work together to provide support and produce movement.

Muscles

There are three major types of muscle in the human body. They're classified by the tissue they contain:

 cardiac (heart) muscle, composed of striated cardiac muscle cells

 smooth (involuntary) muscle, made of smooth muscle cells that lack visible striations

skeletal (voluntary and reflex) muscle, which consists of striated muscle cells.

The attached type

This chapter discusses only skeletal muscle—the type attached to bone. The human body has about 600 skeletal muscles. (See *Viewing the major skeletal muscles.*)

Muscle functions

Skeletal muscles move body parts or the body as a whole. They're responsible for both voluntary and reflex movements. They also maintain posture and generate body heat.

Muscle structure

Skeletal muscle is composed of large, long cells called *muscle fibres*, each of which contains a range of organelles and contractile proteins. Skeletal muscle grows by the fusion of individual cells—which explains why skeletal muscle cells have so many nuclei. (See *Muscle structure up close,* page 40.)

Outside in

The structures of a muscle fibre, working from the cell's exterior to its interior, are:
- *endomysium*—the connective tissue layer surrounding an individual skeletal muscle fibre
- *sarcolemma*—the plasma membrane of the cell that lies beneath the endomysium and just above the cell's nuclei that are pushed to the outside of the cell by contractile proteins
- *sarcoplasm*—the muscle cell's cytoplasm, which is contained within the sarcolemma
- *myofibrils*—consist of myofilaments that run the length of the cell and make up the bulk of the fibre
- *myosin* (thick filaments) and *actin* (thin filaments)—comprise the myofibrils; there are about 1,500 myosin and about 3,000 actin units within each myofibril.

Memory jogger

To remember the functions of muscles, think miles per hour, or **MPH:**
- **M**ovement
- **P**osture
- **H**eat

Body shop

Viewing the major skeletal muscles

This illustration shows anterior and posterior views of some of the major muscles.

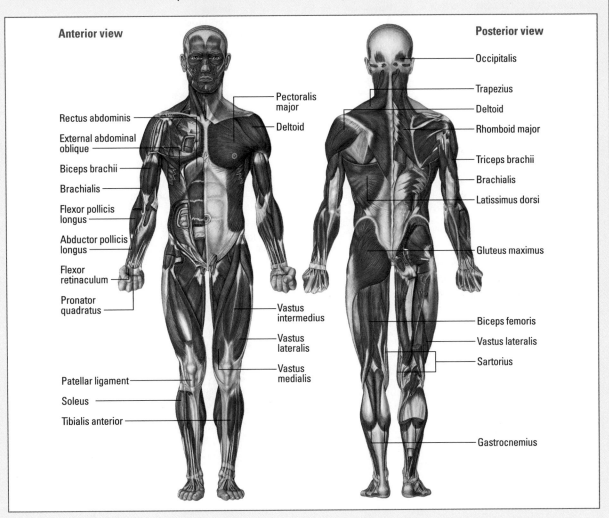

Anterior view

- Rectus abdominis
- External abdominal oblique
- Biceps brachii
- Brachialis
- Flexor pollicis longus
- Abductor pollicis longus
- Flexor retinaculum
- Pronator quadratus
- Patellar ligament
- Soleus
- Tibialis anterior

- Pectoralis major
- Deltoid
- Vastus intermedius
- Vastus lateralis
- Vastus medialis

Posterior view

- Occipitalis
- Trapezius
- Deltoid
- Rhomboid major
- Triceps brachii
- Brachialis
- Latissimus dorsi
- Gluteus maximus
- Biceps femoris
- Vastus lateralis
- Sartorius
- Gastrocnemius

Zoom in

Muscle structure up close

Skeletal muscle contains contractile cells arranged in *fibre bundles* that increase in complexity with the diameter of the muscle tissue. This illustration shows the muscle and its associated fibre bundles.

In a bind

The *perimysium*—a sheath of connective tissue—binds smaller muscle fibres together into an intermediate bundle (fascicle). The *epimysium* binds the fascicles together; beyond the muscle, it becomes a tendon.

Surrounded

A sarcolemma is a thin membrane enclosing a muscle fibre. Myofibrils within the muscle fibres contain even finer structures called *myosin* (thick filaments) and *actin* (thin filaments).

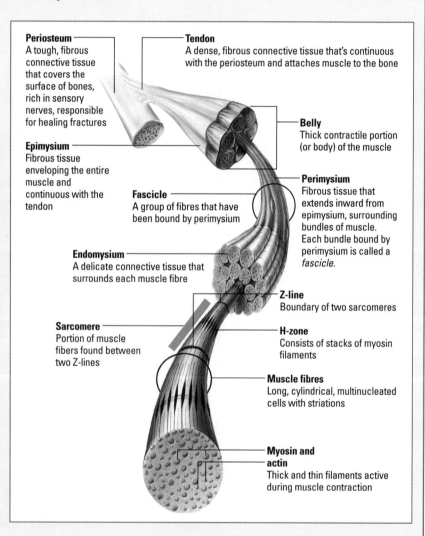

Periosteum
A tough, fibrous connective tissue that covers the surface of bones, rich in sensory nerves, responsible for healing fractures

Tendon
A dense, fibrous connective tissue that's continuous with the periosteum and attaches muscle to the bone

Epimysium
Fibrous tissue enveloping the entire muscle and continuous with the tendon

Belly
Thick contractile portion (or body) of the muscle

Perimysium
Fibrous tissue that extends inward from epimysium, surrounding bundles of muscle. Each bundle bound by perimysium is called a *fascicle.*

Fascicle
A group of fibres that have been bound by perimysium

Endomysium
A delicate connective tissue that surrounds each muscle fibre

Z-line
Boundary of two sarcomeres

H-zone
Consists of stacks of myosin filaments

Sarcomere
Portion of muscle fibers found between two Z-lines

Muscle fibres
Long, cylindrical, multinucleated cells with striations

Myosin and actin
Thick and thin filaments active during muscle contraction

Sarcomeres end to end

Myosin and actin are contained within compartments called *sarcomeres*. Sarcomeres are the functional units of skeletal muscle. During muscle contraction, myosin and actin slide over each other, reducing sarcomere length.

Zoom in

Structure of the sarcomere

The muscle fibre is composed of myofibrils, each of which contains the myofilaments actin and myosin. The regular pattern of overlapping filaments gives skeletal and cardiac muscle their striated appearance. The Z-lines mark the end of the sarcomeres. Within each sarcomere lies a dense area (A-band) which consists mainly of thick filaments (myosin) and portions of the thin filaments (actin) where they overlap the thick filaments. A less dense area (I-band) contains the remainder of the thin filaments. The Z-line passes through the centre of each I-band. A narrow H-zone in the centre of each A-band contains thick but no thin filaments. During contraction, actin filaments slide towards each other increasing the amount of overlap between actin and myosin filaments—the I-band and H-zone decreasing in size as the filaments slide past each other.

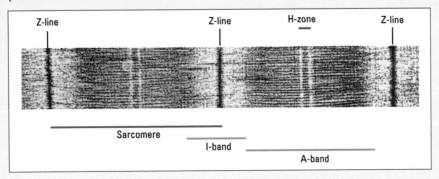

Zebra stripes

The sarcomere compartments of all the myofibrils in a single fibre are aligned. When a muscle fibre is viewed microscopically, transverse (at right angles to the long axis) stripes, called *striations* (or Z-lines), appear along the length of the fibre. Z-lines mark the beginning of sarcomeres. (See *Structure of the sarcomere*.)

A bundle of bundles

A fibrous sheath of connective tissue, called the *perimysium*, binds muscle fibres into a bundle, or *fascicle*. A stronger sheath, the *epimysium*, binds all the fascicles together to form the entire muscle. Extending beyond the muscle, the epimysium becomes a tendon.

Muscle attachment

Most skeletal muscles are attached to bones, either directly or indirectly.

The direct approach

In a direct attachment, the epimysium of the muscle fuses to the *periosteum*, the fibrous membrane covering the bone.

When I contract my muscles to draw my bow, one bone stays stationary. The other bone is pulled towards the stationary one.

Being indirect

In an indirect attachment (most common), the epimysium extends past the muscle as a tendon, or *aponeurosis*, and attaches to the bone.

Contraction

During contraction, one of the bones to which the muscle is attached stays relatively stationary while the other is pulled in towards the stationary one.

Origin and insertion

The point where the muscle attaches to the stationary or less movable bone is called the *origin*; the point where it attaches to the more movable bone is called the *insertion*. The origin usually lies on the proximal end of the bone. The insertion site is on the distal end.

> Factors such as genetic constitution and exercise cause muscle strength and size to differ among individuals.

Muscle growth

Muscle develops when existing muscle fibres *hypertrophy*, largely as a result of an increase in the number of contractile protein units that they contain. Muscle strength and size differ among individuals because of such factors as exercise, nutrition, gender, age, and genetic constitution. Changes in nutrition or exercise affect muscle strength and size in an individual. (See *Musculoskeletal changes with aging*.)

Muscle movements

Skeletal muscle can permit several types of movement. A muscle's functional name comes from the type of movement it permits. For example, a flexor muscle permits bending (*flexion*); an adductor muscle permits movement towards a body axis (*adduction*) and a circumductor muscle allows a circular movement (*circumduction*). (See *Basics of body movement*.)

As time goes by . . .

Musculoskeletal changes with aging

As an individual ages, an apparent musculoskeletal change is decreasing height. This occurs because exaggerated spinal curvature and narrowed intervertebral spaces cause the trunk to shorten and the arms to appear relatively long. This may occur as a result of osteoporosis.

Other musculoskeletal changes that occur with aging include decreased muscle mass, which may result in muscle weakness, and decreased bone density, which causes bones to fracture more readily. In addition, collagen formation declines, which causes joints and supporting structures

to lose resilience and elasticity. Synovial fluid also becomes more viscous and synovial membranes become more fibrotic, making joints stiff.

Ageing may also make tandem walking difficult. Usually, elderly people walk with shorter steps and wider leg stances for better balance and stability.

Body shop

Basics of body movement

Basic muscle movement is best demonstrated in the diarthrodial (freely movable) joints, which allow a large number of angular and circular movements:
- The shoulder demonstrates circumduction.
- The elbow demonstrates flexion and extension.

- The hip demonstrates internal and external rotation.
- The arm demonstrates abduction and adduction.
- The hand demonstrates supination and pronation.
- The jaw demonstrates retraction and protraction.
- The foot demonstrates eversion and inversion.

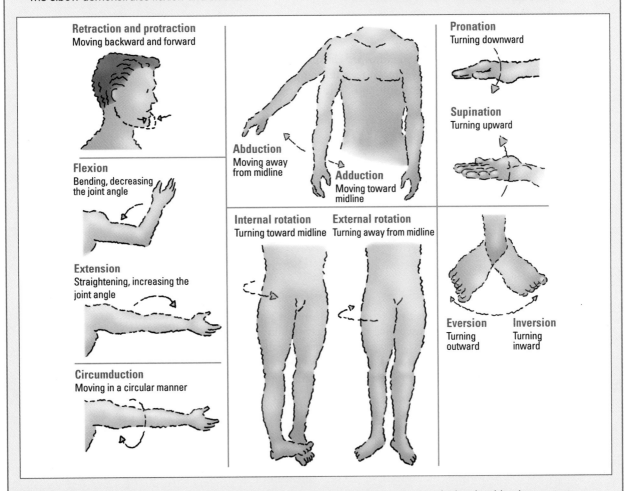

Retraction and protraction
Moving backward and forward

Flexion
Bending, decreasing the joint angle

Extension
Straightening, increasing the joint angle

Circumduction
Moving in a circular manner

Abduction
Moving away from midline

Adduction
Moving toward midline

Internal rotation
Turning toward midline

External rotation
Turning away from midline

Pronation
Turning downward

Supination
Turning upward

Eversion
Turning outward

Inversion
Turning inward

Note that other movements may be attributed to these representative joints. For example the shoulder demonstrates flexion, extension, abduction, adduction, medial and lateral rotation— in addition to circumduction.

Muscles of the axial skeleton

The muscles of the axial skeleton are essential for respiration, speech, facial expression, posture, and chewing. They include:
- muscles of the face, tongue and neck
- muscles of mastication
- muscles of the vertebral column situated along the spine.

Muscles of the appendicular skeleton

The appendicular skeleton includes the muscles of the:
- shoulder
- abdominopelvic cavity
- upper and lower extremities.
    Muscles of the upper extremities are classified according to the bones they move. Those that move the arm are further categorised into those with an origin on the axial skeleton and those with an origin on the scapula.

Tendons and ligaments

Tendons are bands of fibrous connective tissue that attach muscles to the periosteum, the fibrous covering of the bone. Tendons enable bones to move when skeletal muscles contract.
    Ligaments are dense, strong, flexible bands of fibrous connective tissue that attach bones to other bones.

Bones

The human skeleton contains 206 bones: 80 form the *axial skeleton*—called *axial* because it lies along the central line, or axis, of the body—and 126 form the *appendicular skeleton*—relating to the lim or appendages, of the body. (See *Viewing the major bones*.)

Access the axis

Bones of the axial skeleton include:
- facial and cranial bones
- hyoid bone
- vertebrae
- ribs and sternum.

Appendages to the axis

Bones of the appendicular skeleton include:
- clavicle
- scapula
- humerus, radius, ulna, carpals, metacarpals, and phalanges

Appendages are a good thing!

You think you're humerus, don't you?

Body shop

Viewing the major bones

These illustrations show the major bones and bone groups in the body.

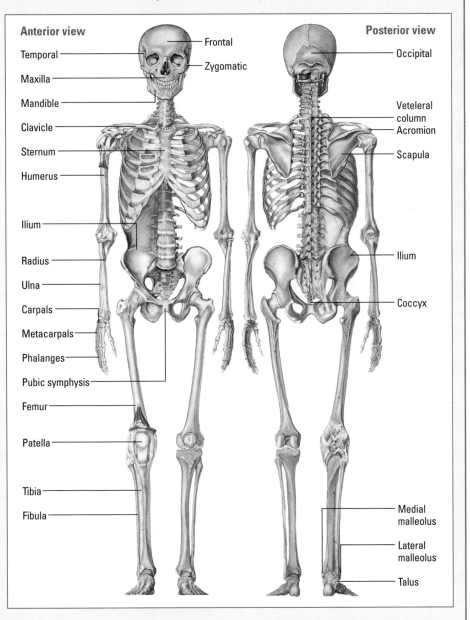

Anterior view

Temporal
Maxilla
Mandible
Clavicle
Sternum
Humerus
Ilium
Radius
Ulna
Carpals
Metacarpals
Phalanges
Pubic symphysis
Femur
Patella
Tibia
Fibula

Frontal
Zygomatic

Posterior view

Occipital
Veteleral column
Acromion
Scapula
Ilium
Coccyx
Medial malleolus
Lateral malleolus
Talus

- pelvic bones
- femur, patella, fibula, tibia, tarsals, metatarsals, and phalanges.

Bone classification

Bones are typically classified by shape. Thus, bones may be classified as:
- long (such as the humerus, radius, femur and tibia) (*See Viewing a long bone.*)
- short (such as the carpals and tarsals)
- flat (such as the scapula, ribs and some of the bones of the skull)
- irregular (such as the vertebrae and mandible)
- sesamoid, which is a small bone developed in a tendon (such as the patella).

Bone functions

Bones perform various anatomical (mechanical) and physiological functions. They:
- protect internal tissues and organs
- stabilise and support the body
- provide a surface for muscle, ligament and tendon attachment
- move through 'lever' action when contracted
- produce blood cells in the bone marrow (*haematopoiesis*)
- store mineral salts (such as 99% of the body's calcium).

I support and stabilise the body.

Blood supply

Blood reaches bones through three paths:

Haversian canals, minute channels that lie parallel to the axis of the bone and are passages for arterioles and other essential structures

Volkmann's canals, which contain vessels that connect one Haversian canal to another and to the outer bone

vessels in the bone ends and in portions of bone containing red marrow.

Bone formation

At 3 months *in utero*, the foetal skeleton is composed of cartilage. By about 6 months, foetal cartilage has been transformed into bony skeleton. (See *Bone growth and remodelling*, page 48.)

Ossification is hard work

By the time of birth, most bones have *ossified* (hardened). The change results from *endochondral ossification*, a process by which *osteoblasts* (bone-forming cells) produce *osteoid* (a collagenous material that ossifies).

(Text continues on page 50.)

Zoom in

Viewing a long bone

The main parts of a long bone, shown below left, are the *diaphysis* (shaft) and the *epiphyses* (ends). The *metaphysis* is the flared end of the diaphysis where the shaft merges with the epiphysis. Periosteum surrounds the diaphysis; endosteum lines the medullary cavity. At the epiphyseal line, cartilage separates the epiphyses from the diaphysis.

Two types of bone tissue

Each bone consists of an outer layer of dense, smooth compact bone, which contains Haversian canals, and an inner layer of spongy cancellous bone, which lacks these canals.

Cancellous bone

Cancellous bone consists of tiny spikes, called *trabeculae*, that interlace to form a latticework. Red marrow fills the spaces between the trabeculae of some bones. Cancellous bone fills the central regions of the epiphyses and the inner portions of short, flat, and irregular bones and also lines the shaft of long bones.

Compact bone

Compact bone is found in the diaphyses of long bones and the outer layers of short, flat and irregular bones. Compact bone consists of layers of calcified matrix containing spaces occupied by osteocytes (mature bone cells). *Lamellae* (bone layers) are arranged around central canals (Haversian canals). Small cavities called *lacunae*, which lie between the lamellae, contain osteocytes. Canaliculi (tiny canals) connect the lacunae, forming the structural units of the bone. Canaliculi allow the movement of nutrients and wastes between the osteocytes.

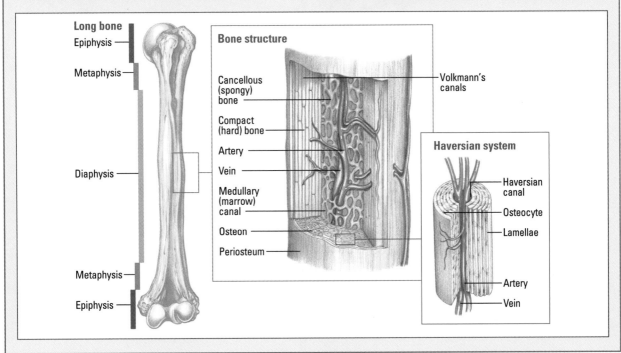

Zoom in

Bone growth and remodelling

The ossification of cartilage into bone, or *osteogenesis*, begins at about the ninth week of foetal development. The diaphyses of long bones are formed by birth, and the epiphyses begin to ossify at about that time. Here are the stages of bone growth and remodelling of the epiphyses of a long bone.

Creation of an ossification centre

At about the ninth month of foetal development, a secondary ossification centre develops in the epiphysis. Some cartilage cells enlarge and stimulate ossification of surrounding cells. The enlarged cells die, leaving small cavities. New cartilage continues to develop.

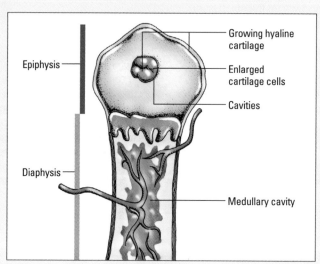

Epiphysis

Diaphysis

Growing hyaline cartilage

Enlarged cartilage cells

Cavities

Medullary cavity

Osteoblasts form bone

Osteoblasts begin to form bone on the remaining cartilage, creating the trabecular network of cancellous bone. Cartilage continues to form on the outer surfaces of the epiphysis and along the upper surface of the epiphyseal plate.

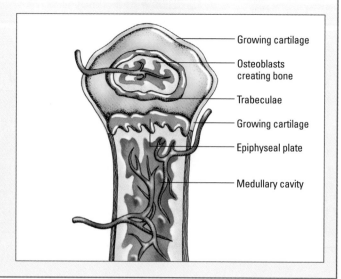

Growing cartilage

Osteoblasts creating bone

Trabeculae

Growing cartilage

Epiphyseal plate

Medullary cavity

Bone growth and remodelling *(continued)*

Bone length grows

Cartilage is replaced by compact bone near the outer surfaces of the epiphysis. Only cartilage cells on the upper surface of the epiphyseal plate continue to multiply rapidly, pushing the epiphysis away from the diaphysis. This new cartilage ossifies, creating trabeculae on the medullary side of the epiphyseal plate.

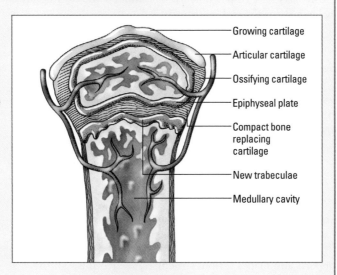

- Growing cartilage
- Articular cartilage
- Ossifying cartilage
- Epiphyseal plate
- Compact bone replacing cartilage
- New trabeculae
- Medullary cavity

Remodelling

Osteoclasts produce enzymes and acids that reduce trabeculae created by the epiphyseal plate, thus enlarging the medullary cavity. In the epiphysis, osteoclasts reduce bone, making its calcium available for new osteoblasts that give the epiphysis its adult shape and proportion. Around age 25, the epiphyseal plate completely ossifies (closes) and becomes the epiphyseal line; longitudinal growth of bone then ceases.

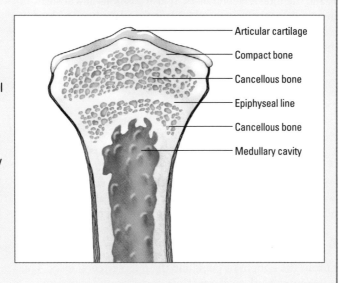

- Articular cartilage
- Compact bone
- Cancellous bone
- Epiphyseal line
- Cancellous bone
- Medullary cavity

Bone density decreases after age 40 in women and after age 50 in men.

Bone remodelling

Osteoblasts and osteoclasts are responsible for *remodelling*—the continuous process whereby bone is created and destroyed.

Blast these bones!

Osteoblasts are responsible for the formation of new bone and *osteoclasts* are involved in the reabsorption of calcified bone. Together these cells act to promote longitudinal bone growth. This growth continues until the *epiphyseal plates* ossify during late adolescence. The epiphyseal plates are cartilage that separate the *diaphysis*, or shaft of a bone, from the *epiphysis*, or end of a bone.

Cartilage

Cartilage is a dense connective tissue that consists of fibres embedded in a strong, gel-like substance. Unlike rigid bone, cartilage has the flexibility of firm plastic.

Cartilage supports and shapes various structures, such as the auditory canal, the larynx and the intervertebral discs. It also cushions and absorbs shock, preventing direct transmission to the bone. Cartilage has no blood supply or innervation.

There are three types of cartilage:

 hyaline

 fibrous

 elastic.

Common connector

Hyaline cartilage is the most common type of cartilage. It covers the articular bone surfaces (where one or more bones meet at a joint). It also connects the ribs to the sternum and appears in the trachea, bronchi and nasal septum.

The strong, rigid type

Fibrous cartilage forms the symphysis pubis and the intervertebral discs. This type of cartilage is composed of small quantities of matrix and abundant fibrous elements. It's strong and rigid.

Staying flexible

Elastic cartilage, the most pliable cartilage, is located in the auditory canal, external ear and epiglottis. Large numbers of elastic fibres give this type of cartilage elasticity and resiliency.

Joints

Joints (articulations) are points of contact between two bones, which hold the bones together. Many joints also allow flexibility and movement.

Joint classification

Joints can be classified by function (extent of movement) or by structure (what they're made of). The body has three major types of joints classified by function and three major types classified by structure.

Functional classification

By function, a joint may be classified as:

 synarthrosis (immovable)

 amphiarthrosis (partially movable)

 diarthrosis (freely movable).

Structural classification

By structure, a joint may be classified as:

 fibrous

 cartilaginous

 synovial.

Fibrous joints

With *fibrous joints*, the articular surfaces of the two bones are bound closely by fibrous connective tissue, and little movement is possible. Fibrous joints include *sutures*, *syndesmoses* (such as the radioulnar joints) and *gomphoses* (such as the dental alveolar joint).

Cartilaginous joints

With *cartilaginous joints* (also called amphiarthroses), cartilage connects one bone to another. Cartilaginous joints allow slight movement. They occur as:
• *synchondroses*, which are typically temporary joints in which the intervening hyaline cartilage converts to bone by adulthood—for example, the epiphyseal plates of long bones
• *symphyses*, which are joints with an intervening pad of fibrocartilage—for example, the symphysis pubis.

Hey, I know a joint where we can get together.

Synovial joints

The continuous bony surfaces in the *synovial joints* are separated by a viscous lubricating fluid (*synovial fluid*) and by cartilage. They're joined by ligaments lined with a membrane that produces the synovial fluid. Freely movable or diarthrosis, synovial joints include most joints of the arms and legs.

Other features of synovial joints include:
- a *joint cavity*—a potential space that separates the articulating surfaces of the two bones
- an *articular capsule*—a saclike envelope with an outer layer that's lined with a vascular synovial membrane
- *reinforcing ligaments*—fibrous tissue that connects bones within the joint and reinforces the joint capsule.

Joint subdivisions

Based on their structure and the type of movement they allow, synovial joints fall into various subdivisions—gliding, hinge, pivot, condylar, saddle and ball-and-socket. (See *Types of synovial joint*.)

Let it glide

Gliding joints have flat or slightly curved articular surfaces and allow gliding movements. However, because they're bound by ligaments, they may not allow movement in all directions. Examples of gliding joints are the intertarsal and intercarpal joints of the hands and feet.

Here's a hinge

With *hinge joints*, a convex portion of one bone fits into a concave portion of another. The movement of a hinge joint resembles that of a metal hinge and is limited to flexion and extension. Hinge joints include the elbow and knee.

Synovial joints are freely movable and include most joints of the arms and legs.

Bones are like levers; joints are like fulcrums... that's a pivot around which a lever moves.

Apply your muscles and the lever moves. Wheeeee!

Zoom in

Types of synovial joint

Synovial joints (diarthroses) have a distinguishing characteristic (the synovial cavity) that separates the articulating bones. Another characteristic of such bones is the presence of *articular cartilage*. Articular cartilage covers the surfaces of the articulating bones but does not bind the bones together. This cartilage reduces friction when bones move and helps absorb shock. There are several subtypes of synovial joint. These are illustrated below, together with an indication of some of the locations in which they may be found and the type of movement they afford.

a. gliding joints
- found in the carpals of the wrist and the acromioclavicular joint, these allow only gliding or sliding movements

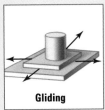

Gliding

b. hinge joints
- found at the elbow (between the humerus and the ulna), these joints allow flexion and extension in just one plane

Hinge

c. pivot joints
- found at the atlantoaxial, proximal radioulnar and distal radioulnar joints allowing one bone to rotate around another

Pivot

d. condylar joints
- found at the joint of the wrist (radiocarpal joint) these articulations allow bones that have a concave and a convex shape to fit together, allowing movement in two planes

Condylar

e. saddle joints
- found uniquely in the carpometacarpal joint of each thumb, these joints permit the same movements as condylar joints

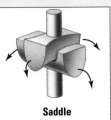

Saddle

f. ball-and-socket joints
- found at the shoulder (glenohumeral) and hip joints, these allow a wide range of movement

Ball and Socket

'Pivotal' joints

A rounded portion of one bone in a *pivot joint* fits into a groove in another bone. Pivot joints allow only uniaxial rotation of the first bone around the second. An example of a pivot joint is the head of the radius, which rotates within a groove of the ulna.

Give condylar joints a hand

With *condylar joints*, an oval surface of one bone fits into a concavity in another bone. Condylar joints allow flexion, extension, abduction, adduction and circumduction. Examples include the radiocarpal and metacarpophalangeal joints of the hand.

Saddle up

Saddle joints resemble condylar joints but allow greater freedom of movement. The only saddle joints in the body are the carpometacarpal joints of the thumbs.

Ball-and-socket: these joints are hip

The *ball-and-socket joint* gets its name from the way its bones connect: the spherical head of one bone fits into a concave 'socket' of another bone. The body's only ball-and-socket joints are the shoulder and hip joints.

Hey, stranger. The only saddle joints in the body are the carpometacarpal joints of the thumb.

Bursae

Bursae are small synovial fluid sacs that are located at friction points around joints between tendons, ligaments and bones.

Stress reducers

Bursae act as cushions to decrease stress on adjacent structures. Examples of bursae include the subacromial bursa (located in the shoulder) and the prepatellar bursa (a superficial bursa with a thin synovial lining located between the skin and the patella).

Quick quiz

1. Which of these features is characteristic of a smooth muscle cell?
   A. Branched appearance
   B. Multinucleated
   C. Spindle shaped
   D. Voluntary

Answer: C. Smooth muscle is typically spindle shaped. Smooth muscle cells contain a single nucleus and are involuntary (controlled by the autonomic nervous system). They are generally located in the walls of hollow internal organs.

2. Which statement is true about cartilage?
   A. It receives a generous blood supply.
   B. It protects body structures.
   C. It is completely flexible.
   D. It cushions and absorbs shock.

Answer: D. Cartilage is responsible for supporting, cushioning and shaping body structures. Types of cartilage include fibrous, hyaline and elastic.

3. The type of joint that permits free movement is classified as:
   A. Synarthrosis.
   B. Cartilaginous.
   C. Diarthrosis.
   D. Fibrous.

Answer: C. Diarthroses include the ankles, wrists, knees, hips and shoulders. These joints permit free movement.

4. The carpometacarpal joints of the thumb are classified as:
   A. Pivot joints.
   B. Saddle joints.
   C. Hinge joints.
   D. Gliding joints.

Answer: B. Saddle joints are similar to condylar joints. The only saddle joints in the body are the carpometacarpal joints of the thumbs.

Scoring

☆☆☆   If you answered all four questions correctly, outstanding! You've flexed your mental muscles and are ready to tackle the next system.

☆☆   If you answered three questions correctly, splendid! You've got bones of this system.

☆   If you answered fewer than three questions correctly, that's okay. Let's take our osteocytes and blast through this chapter one more time.

Just the facts

In this chapter, you'll learn:

♦ structures of the nervous system

♦ functions of the nervous system

♦ special sense organs and their functions.

A look at the neurosensory system

The nervous system coordinates all body functions, enabling a person to adapt to changes in internal and external environments. It has two main types of cells:

- neurones, the conducting cells
- neuroglia, the supporting cells.

It comprises two major divisions, the *central nervous system (CNS)* consisting of the brain and spinal cord and the *peripheral nervous system (PNS)* consisting of the cranial and spinal nerves; the PNS has somatic and autonomic components. (See *Organisation of the nervous system*.)

Neurone: the basic unit

The *neurone* is the basic unit of the nervous system. This highly specialised conductor cell receives and transmits electrochemical nerve impulses. Delicate, nerve fibres called *axons* and *dendrites* extend from the central cell body and transmit signals. In a typical neurone, one axon and many dendrites extend from the cell body. (See *Parts of a neurone*, page 58.)

Axons

Axons conduct nerve impulses away from cell bodies. A typical axon has terminal branches and is wrapped in a white, fatty, segmented covering called a *myelin sheath*. The myelin sheath is produced by *Schwann cells*—neuroglial cells separated by gaps called *nodes of Ranvier*.

I'm a neurone, the fundamental unit of the nervous system.

Now I get it!

Organisation of the nervous system

The nervous system is divided primarily into the central nervous system (CNS) and the peripheral nervous system (PNS). The CNS consists of the brain and spinal cord. Within the CNS incoming sensory information is integrated and correlated; this includes thought processes, feelings of emotion and formation of memory. Most nerve impulses that stimulate muscles and glands originate in the CNS. The CNS is connected to sensory receptors, muscles and glands in the outlying parts of the body, by the PNS, which consists of sensory input (afferent) and motor output (efferent) neurones.

The PNS may be subdivided further into somatic and autonomic components. The somatic component (SNS) consists of sensory neurones that convey information from the cutaneous and special sense receptors to the CNS and motor neurones from the CNS that conduct impulses solely to skeletal muscle (voluntary control). The autonomic component (autonomic nervous system, ANS) consists of sensory neurones that convey information from receptors primarily in the viscera to the CNS and motor neurones from the CNS that conduct impulses to smooth muscle, cardiac muscle and glands.

The motor division of the ANS consists of the sympathetic and parasympathetic branches. With a few exceptions visceral tissues receive instructions from both. The two divisions usually have opposing actions.

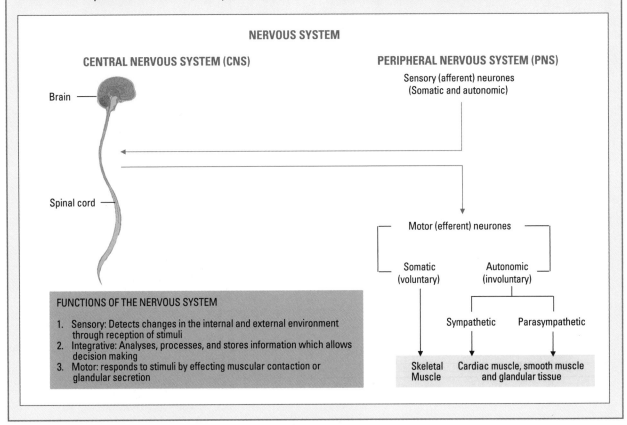

NERVOUS SYSTEM

CENTRAL NERVOUS SYSTEM (CNS)

PERIPHERAL NERVOUS SYSTEM (PNS)

Sensory (afferent) neurones
(Somatic and autonomic)

Brain

Spinal cord

Motor (efferent) neurones

Somatic
(voluntary)

Autonomic
(involuntary)

Sympathetic   Parasympathetic

Skeletal
Muscle

Cardiac muscle, smooth muscle
and glandular tissue

FUNCTIONS OF THE NERVOUS SYSTEM

1. Sensory: Detects changes in the internal and external environment through reception of stimuli
2. Integrative: Analyses, processes, and stores information which allows decision making
3. Motor: responds to stimuli by effecting muscular contaction or glandular secretion

Zoom in

Parts of a neurone

A typical motor neurone, like the one shown here, has one axon and many dendrites. A myelin sheath encloses the axon.

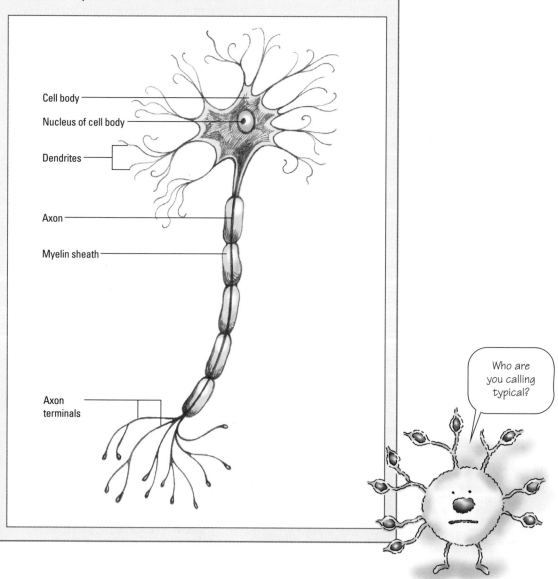

Cell body

Nucleus of cell body

Dendrites

Axon

Myelin sheath

Axon terminals

Who are you calling typical?

Dendrites

Typically, *dendrites* are diffusely branched extensions of the cell body that receive impulses from other cells. Dendrites conduct impulses towards the cell body.

Sending the message

Neurones are responsible for *neurotransmission*— conduction of electrochemical impulses throughout the nervous system. A range of specialised receptor cells allow individual neurones to respond to specific stimuli. Neuronal activity may be provoked by:

• mechanical stimuli, such as touch and pressure
• thermal stimuli, such as heat and cold
• chemical stimuli, such as external chemicals or a chemical released by the body, such as histamine. (See *How neurotransmission occurs*, page 60.)

> Thermal stimuli such as heat produce neurone activity and start conduction of electrochemical impulses through the nervous system.

The reflex arc

The reflex arc—a neural relay cycle for quick motor response to a harmful sensory stimuli—requires a sensory (afferent) neurone and a motor (efferent) neurone. The stimulus triggers a sensory impulse, which travels along the dorsal root to the spinal cord. There, two synaptic transmissions occur at the same time. One synapse continues the impulse along a sensory neurone to the brain; the other immediately relays the impulse to an interneurone, which transmits it to a motor neurone. (See *The reflex arc*, page 61.)

Neuroglia

Neuroglia (also called *glial cells*) are the supporting cells of the nervous system. They form roughly 40% of the brain's bulk.

> Glial is derived from the Greek word for glue. Glial cells 'glue' the neurones together.

Neuroglia to know

There are six types of neuroglia:

Astroglia, or *astrocytes*, exist throughout the nervous system. They supply nutrients to neurones and help them maintain their electrical potential. They also form part of the blood-brain barrier, which prevents some harmful molecules from entering the brain.

Ependymal cells line the four small cavities in the brain, called *ventricles*, and the choroid plexuses. They help produce and circulate cerebrospinal fluid (CSF).

Microglia are phagocytic cells that ingest and digest microorganisms and waste products from injured neurones.

Oligodendroglia support and electrically insulate CNS axons by forming protective myelin sheaths.

Now I get it!

How neurotransmission occurs

Neurones receive and transmit stimuli by electrochemical messages. Neurones may be directly stimulated or dendrites on the neurone may receive an impulse sent by other cells and conduct it towards the cell body. The axon then conducts the impulse away from the cell.

To stimulate or inhibit

When the impulse reaches the end of the axon, it stimulates synaptic vesicles in the presynaptic axon terminal. A neurotransmitter substance is then released into the synaptic cleft between neurones. This substance diffuses across the synaptic cleft and binds to specific receptors molecules on the postsynaptic membrane. This stimulates or inhibits activity of the postsynaptic neurone.

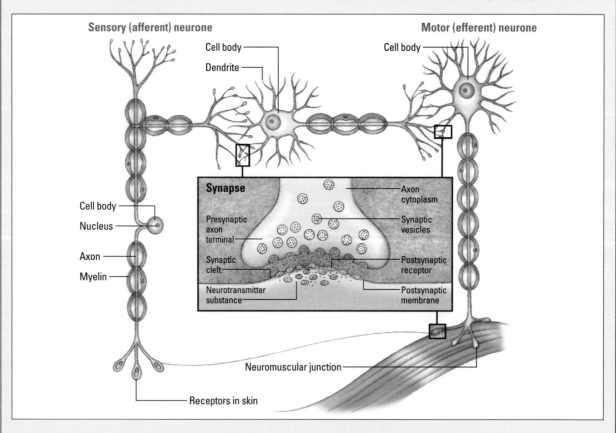

Sensory (afferent) neurone

Motor (efferent) neurone

Cell body

Dendrite

Cell body

Cell body

Nucleus

Axon

Myelin

Synapse

Presynaptic axon terminal

Synaptic cleft

Neurotransmitter substance

Axon cytoplasm

Synaptic vesicles

Postsynaptic receptor

Postsynaptic membrane

Neuromuscular junction

Receptors in skin

Now I get it!

The reflex arc

The reflex arc is the transmission of sensory impulses to a motor neurone via the dorsal root. The motor neurone delivers the impulse to a muscle or gland, producing an immediate response. Descending motor pathways may initiate a protective response such as cooling the finger under cold running water.

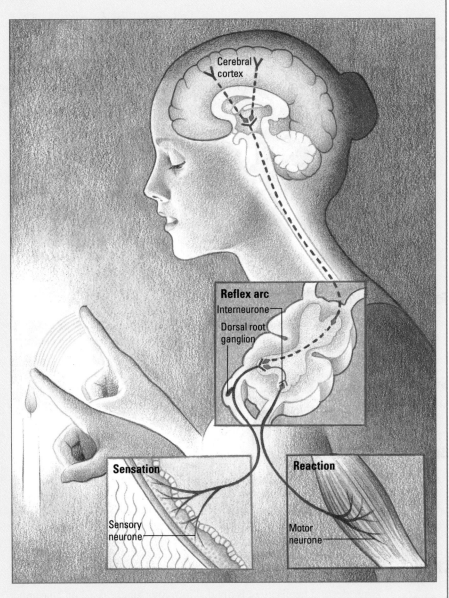

Neurolemmocytes (or Schwann cells) produce myelin sheaths around PNS neurons.

Satellite cells in the PNS serve a similar function to astrocytes in the CNS.

Central nervous system

The CNS includes the brain and the spinal cord. Encased by the bones of the skull and vertebral column, the CNS is protected by the CSF and the meninges (the dura mater, arachnoid and pia mater).

Brain

The brain consists of the cerebrum, cerebellum, brain stem, diencephalon (thalamus and hypothalamus), limbic system and reticular activating system (RAS).

Cerebrum

The *cerebrum* is the largest part of the brain. It houses the nerve centre that controls sensory and motor activities and intelligence.

Touch of grey

The outer layer of the cerebrum, the *cerebral cortex*, consists of unmyelinated nerve fibres *(grey matter)*. The inner layer of the cerebrum consists of myelinated nerve fibres *(white matter)*.

Steady as she goes

Basal nuclei, which control motor coordination and steadiness, are found in white matter.

Bridging the hemispheres

The cerebrum has right and left hemispheres. A mass of nerve fibres known as the *corpus callosum* bridges the hemispheres, allowing communication between corresponding centres in each hemisphere. The rolling surface of the cerebrum is made up of *gyri* (convolutions) and *sulci* (creases) and deep fissures.

The four lobes

Each cerebral hemisphere is divided into four lobes, based on anatomic landmarks and functional differences. These lobes—the frontal, temporal, parietal and occipital—are named after the cranial bones that lie over them. (See *A close look at major brain structures.*)

Basal nuclei, which control motor coordination and steadiness, are found as pockets of grey in white matter.

Zoom in

A close look at major brain structures

The illustration below shows the two largest structures of the brain—the cerebrum and cerebellum. Several fissures divide the cerebrum into hemispheres and lobes:

- The lateral sulcus separates the temporal lobe from the frontal and parietal lobes.
- The central sulcus separates the frontal lobes from the parietal lobe.
- The *parieto-occipital fissure* separates the occipital lobe from the two parietal lobes.

To each lobe, a function

Each lobe has a particular function:

- The *frontal lobes* influence personality, judgment, abstract reasoning, social behaviour, language expression and voluntary movement (in the motor portion).
- The *temporal lobes* control hearing, language comprehension, learning, understanding and storage and recall of memories (although memories are stored throughout the entire brain).
- The *parietal lobes* interpret and integrate sensations, including pain, temperature and touch. They also interpret size, shape, distance, vibration and texture. The parietal lobe of the nondominant hemisphere is especially important for awareness of body shape.
- The *occipital lobes* function mainly to interpret visual stimuli.

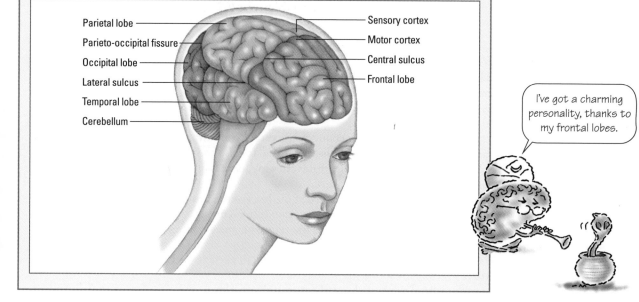

Parietal lobe

Parieto-occipital fissure

Occipital lobe

Lateral sulcus

Temporal lobe

Cerebellum

Sensory cortex

Motor cortex

Central sulcus

Frontal lobe

I've got a charming personality, thanks to my frontal lobes.

Cerebellum

The *cerebellum* is the brain's second largest region. It lies behind and below the cerebrum. Like the cerebrum, it has two hemispheres. It also has an outer cortex of grey matter and an inner core of white matter. The cerebellum functions to maintain muscle tone, coordinate muscle movement and control balance.

Brain stem

The *brain stem* lies immediately below the cerebrum, just in front of the cerebellum. It is continuous with the cerebrum above and connects with the spinal cord below.

Thanks to the brain stem, I can communicate with the rest of the nervous system.

It all stems from the brain stem

The brain stem consists of the *midbrain, pons* and *medulla oblongata*. It relays messages between the parts of the nervous system and has three main functions:
• It produces the vital autonomic reactions necessary for survival, such as increasing heart rate and stimulating the adrenal medulla to produce adrenaline (epinephrine).
• It mediates the sneeze reflex.
• It provides pathways for nerve fibres between higher and lower neural centres.
• It serves as the origin for 10 of the 12 pairs of cranial nerves (CNs).

It goes both ways

The three parts of the brain stem provide two-way conduction between the spinal cord and brain. In addition, they perform the following functions:
• The *midbrain* is the reflex centre for CNs III and IV and mediates pupillary reflexes and eye movements. It also relays auditory information from the ears to the cerebral cortex.
• The *pons* helps regulate respirations. It connects the cerebellum with the cerebrum and links the midbrain to the medulla oblongata. It's also the reflex centre for CNs V through VIII. The pons mediates chewing, taste, saliva secretion, hearing and equilibrium.
• The *medulla oblongata* becomes the spinal cord at the level of the *foramen magnum*, an opening in the occipital portion of the skull. It influences cardiac, respiratory and vasomotor functions. It's the centre for the vomiting, coughing and hiccuping reflexes.

Diencephalon

The *diencephalon* is the part of the brain located between the cerebrum and the midbrain. It consists of the thalamus, epithalamus and hypothalamus, which lie deep in the cerebral hemispheres.

Screening calls

The *thalamus* relays sensory stimuli as they ascend to the cerebral cortex. It also transmits olfactory information to the temporal lobes and limbic system. Its functions include primitive awareness of pain, screening of incoming stimuli and focusing of attention. It also has a role in learning and memory.

Control centre

The *hypothalamus* controls or affects body temperature, appetite, water balance, pituitary secretions, emotions and autonomic functions, including circadian cycles.

Limbic system

The *limbic system* is a primitive brain area deep within the temporal lobe. In addition to initiating basic drives (such as hunger, aggression and emotional and sexual arousal), the limbic system screens many sensory messages traveling to the cerebral cortex. (See *Parts of the limbic system*.)

My thalamus acts as a relay station. Sensory impulses travel across synapses in the thalamus on their way to the cerebral cortex.

Zoom in

Parts of the limbic system

The illustration below shows the structures of the limbic system.

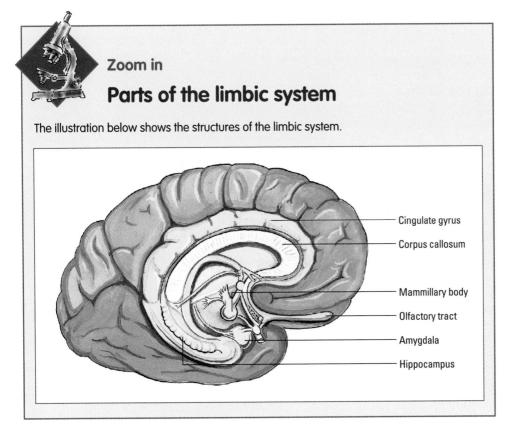

- Cingulate gyrus
- Corpus callosum
- Mammillary body
- Olfactory tract
- Amygdala
- Hippocampus

As time goes by . . .

Neurological changes with aging

Aging affects the nervous system in many ways. For example, neurones of the central and peripheral nervous systems undergo degenerative changes. After about age 50, the number of brain cells decreases by about 1% per year. However, clinical effects usually aren't noticeable until well into old age.

As a person ages, the hypothalamus becomes less effective at regulating body temperature. Also, the cerebral cortex undergoes a 20% neurone loss in later years. Because nerve transmission typically slows down, the older adult may react sluggishly to external stimuli.

Reticular activating system

The *reticular activating system* (RAS) is a diffuse network of hyperexcitable neurones. It fans out from the brain stem through the cerebral cortex. After screening all incoming sensory information, the RAS channels it to appropriate areas of the brain for interpretation. It functions as the arousal, or alerting, system for the cerebral cortex and is crucial in maintaining consciousness. (See *Neurological changes with aging*.)

Oxygenating the brain

Four major arteries—two vertebral and two carotid—supply the brain with oxygenated blood.

Vertebral convergence

The two *vertebral arteries* (branches of the subclavians) converge to become the basilar artery. The *basilar artery* supplies blood to the posterior brain.

Two carotids diverged in the brain . . .

The common carotids branch into the two internal carotids, which divide further to supply blood to the anterior brain and the middle brain.

These arteries interconnect through the cerebral arterial circle (*circle of Willis*), an anastomosis at the base of the brain. The cerebral arterial circle ensures that blood continually circulates to the brain despite interruption of any of the brain's major vessels. (See *Arteries of the brain*.)

Because of the cerebral arterial circle, blood has two paths to the brain, ensuring a continuous blood supply.

Zoom in

Arteries of the brain

This illustration shows the inferior surface of the brain. The anterior and posterior arteries join with smaller arteries to form the circle of Willis.

Inferior view

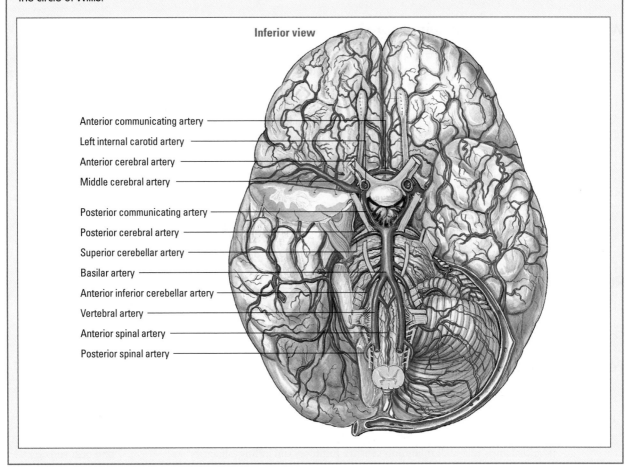

Anterior communicating artery

Left internal carotid artery

Anterior cerebral artery

Middle cerebral artery

Posterior communicating artery

Posterior cerebral artery

Superior cerebellar artery

Basilar artery

Anterior inferior cerebellar artery

Vertebral artery

Anterior spinal artery

Posterior spinal artery

Spinal cord

The *spinal cord* is a cylindrical structure in the vertebral canal that extends from the medulla oblongata exiting the brain at the foramen magnum.

Getting on my spinal nerves

The *spinal nerves* arise from the cord. At the cord's inferior end, nerve roots cluster in the *cauda equina*.

What's the matter in the spinal cord?

Within the spinal cord, the H-shaped mass of grey matter is divided into *horns*. Horns consist mainly of neurone cell bodies. Cell bodies in the two dorsal (posterior) horns primarily relay sensory impulses to the sensory cortex in the brain; those in the two ventral (anterior) horns play a part in voluntary and reflex motor activity. White matter surrounds the horns. This white matter consists of myelinated nerve fibres grouped in vertical columns, or *tracts*. In other words, all axons that compose one tract serve one general function, such as touch, movement, pain and pressure. (See *A look inside the spinal cord*.)

Sensory pathways

Sensory impulses travel via the *afferent* (sensory, or ascending) *neural pathways* to the *sensory cortex* in the parietal lobe of the brain. This is where the impulses are interpreted. They use four pairs of sensory tracts.

Running hot and cold

Pain and temperature sensations enter the spinal cord through the *dorsal horn*. After immediately crossing over to the opposite side of the cord, these impulses then travel to the thalamus via the lateral spinothalamic tract.

Feeling the pressure

Touch, pressure, vibration and pain sensations enter the cord via relay stations called *ganglia*. Ganglia (cell bodies of sensory neurons) lie in the dorsal roots of spinal nerves. Impulses travel up the cord in the dorsal column to the medulla, where they cross to the opposite side and enter the thalamus. The thalamus relays all incoming sensory impulses (including olfactory impulses) to the sensory cortex for interpretation.

Memory jogger

To remember the difference between afferent and efferent neurones, consider this:
- **Afferent** neurones cause sensation to **a**scend to the brain. (Afferent neurones are sensory and ascending.)
- **Efferent** neurones send impulses out of the brain to **e**ffect action. (Efferent neurones are motor and descending.)

Zoom in

A look inside the spinal cord

This cross section of the spinal cord shows an H-shaped mass of grey matter divided into horns, which consist primarily of neurone cell bodies. Cell bodies in the posterior, or dorsal, horn primarily receive information which is relayed by the tracts. Cell bodies in the anterior, or ventral, horn are needed for voluntary or reflex motor activity.

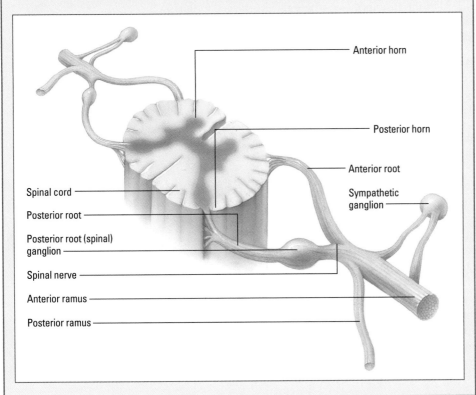

Anterior horn

Posterior horn

Anterior root

Spinal cord

Sympathetic ganglion

Posterior root

Posterior root (spinal) ganglion

Spinal nerve

Anterior ramus

Posterior ramus

Motor tracts

Motor impulses travel from the brain to the muscles via the *efferent* (motor, or descending) *neural pathways*. Motor impulses originate in the *motor cortex* of the frontal lobes and reach the lower motor neurones of the peripheral nervous system via upper motor neurones.

Upper motor neurones originate in the brain and form two major systems:
- the pyramidal system
- the extrapyramidal system.

Fine tuning your response

The *pyramidal system* is responsible for fine, skilled movements of skeletal muscle. Impulses in this system travel from the motor cortex through to the medulla. At the medulla, they cross to the opposite side and continue down the spinal cord.

Get your motor running

The *extrapyramidal system* controls gross motor movements. Impulses originate in the premotor area of the frontal lobes and travel to the pons. At the pons, the impulses cross to the opposite side. Then the impulses travel down the spinal cord to the anterior horn, where they're relayed to the lower motor neurones. These neurones, in turn, carry the impulses to the muscles. (See *Major neural pathways*.)

Reflex responses

Reflex responses occur automatically, without immediate brain involvement, to protect the body. Spinal nerves, which have both sensory and motor neurones, mediate *deep tendon reflexes* (involuntary contractions of a muscle after brief stretching caused by tendon percussion), *superficial reflexes* (withdrawal reflexes elicited by noxious or tactile stimulation of the skin or mucous membranes) and, in infants, *primitive reflexes*.

Deep we go

Deep tendon reflexes include reflex responses of the biceps, triceps, brachioradialis, patellar and Achilles tendons:
- The *biceps reflex* contracts the biceps muscle and forces flexion of the forearm.
- The *triceps reflex* contracts the triceps muscle and forces extension of the forearm.
- The *brachioradialis reflex* causes supination of the hand and flexion of the forearm at the elbow.
- The *patellar reflex* forces contraction of the quadriceps muscle in the thigh with extension of the leg.
- The *Achilles reflex* forces plantar flexion of the foot at the ankle.
(See *Eliciting deep tendon reflexes*, page 72.)

Rising to the superficial

Superficial reflexes are reflexes of the skin and mucous membranes. Successive attempts to stimulate these reflexes provoke increasingly limited responses. Here's a description of some superficial reflexes:
- *Plantar flexion* of the toes occurs when the lateral sole of an adult's foot is stroked from heel to great toe with a tongue depressor.
- *Babinski's reflex* is an upward movement of the great toe and fanning of the little toes that occurs in children under age 2 in response to stimulation of the outer margin of the sole of the foot. This reflex is an abnormal finding in adults.

The pyramidal system is responsible for fine, skilled movements.

Now I get it!

Major neural pathways

Sensory and motor impulses travel through different pathways to and from the brain for interpretation.

Sensory pathways

Sensory impulses travel through two major sensory (afferent, or ascending) pathways to the sensory cortex in the cerebrum.

Motor pathways

Motor impulses travel from the motor cortex in the cerebrum to the muscles via motor (efferent or descending) pathways.

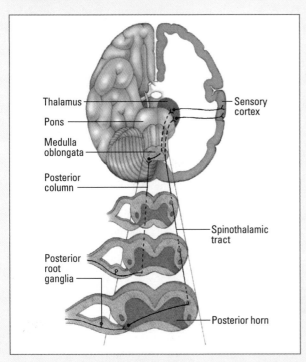

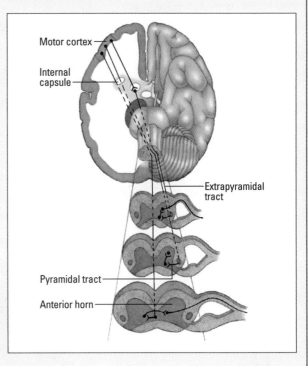

• In males, the *cremasteric reflex* is stimulated by stroking the inner thigh. This forces the contraction of the cremaster muscle and elevation of the testicle on the side of the stimulus.

• The *abdominal reflexes* are induced by stroking the sides of the abdomen above and below the umbilicus, moving from the periphery towards the midline. Movement of the umbilicus towards the stimulus is normal.

• The *startle reflex* is normally present in all newborns and infants up to 4 or 5 months of age, and its absence indicates a profound disorder of the motor system. Persistence of this reflex response beyond 4 or 5 months of age is noted only in infants with severe neurological defects.

Eliciting deep tendon reflexes

There are five deep tendon reflexes. Methods for eliciting these reflexes are described below.

Biceps reflex

Placing the thumb or index finger over the biceps tendon and the remaining fingers loosely over the triceps muscle, strike the thumb or index finger over the biceps tendon with the pointed end of the reflex hammer. Watch and feel for the contraction of the biceps muscle and flexion of the forearm.

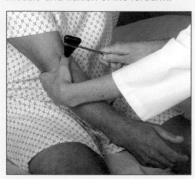

Triceps reflex

Strike the triceps tendon about 5 cm above the olecranon process on the extensor surface of the upper arm. Watch for contraction of the triceps muscle and extension of the forearm.

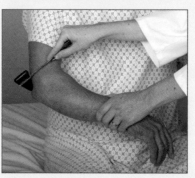

Brachioradialis reflex

Strike the radius about 2.5 to 5 cm above the wrist and watch for supination of the hand and flexion of the forearm at the elbow.

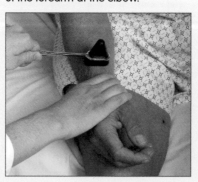

Patellar reflex

Strike the patellar tendon just below the patella and look for contraction of the quadriceps muscle in the thigh with extension of the leg.

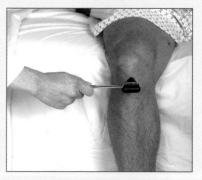

Achilles reflex

With the foot flexed and supporting the plantar surface, strike the Achilles tendon. Watch for plantar flexion of the foot at the ankle.

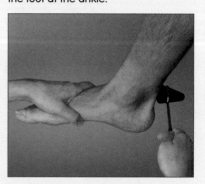

Deep tendon reflexes help protect the body.

Let's get primitive

Primitive reflexes are abnormal in adults but normal in infants, whose central nervous systems are immature. As the neurological system matures, these reflexes disappear. The primitive reflexes are *grasping*, *sucking* and *glabella*:
- The application of gentle pressure to an infant's palm results in grasping.
- An infantile sucking reflex to ingest milk is a primitive response to oral stimuli.
- The glabella reflex is elicited by repeatedly tapping the infant on the bridge of the nose or between the eyebrows. The normal response is persistent blinking.

Protective structures

The brain and spinal cord are protected from shock and the possibility of infection by the bony skull and vertebrae, CSF and three membranes: the dura mater, arachnoid membrane and pia mater. Together these constitute the meninges.

Hey, we all need a little protection sometimes!

Dura mater
The *dura mater* is tough, fibrous, leatherlike tissue composed of two layers—the endosteal dura and meningeal dura.

The unending endosteal dura

The *endosteal dura* forms the periosteum of the skull and is continuous with the lining of the vertebral canal.

The durable meningeal dura

The *meningeal dura* is a thick membrane that covers the brain, dipping between the brain tissue and providing support and protection. It lies deep to the endosteal dura.

Arachnoid membrane
The *arachnoid membrane* is a thin, fibrous membrane that hugs the brain and spinal cord, though not as precisely as the pia mater.

Pia mater
The *pia mater* is a continuous, delicate layer of connective tissue that covers and contours the spinal tissue and brain.

The spaces between

The *subdural space* lies between the dura mater and the arachnoid membrane. The *subarachnoid space* lies between the arachnoid membrane and the pia mater. Within the subarachnoid space and the brain's four ventricles is CSF, a plasma derived fluid composed of water and traces of organic materials (including protein), glucose and electrolytes. This fluid protects the brain and spinal tissue from jolts and blows. CSF also provides nutrients and various chemicals such as neurotransmitters and hormones to brain cells. The fluid may be sampled for biochemical, microbiological and cytological analysis by means of a *lumbar puncture*.

Peripheral nervous system

The PNS consists of the CNs, spinal nerves and autonomic nervous system (ANS).

Cranial nerves

Twelve pairs of CNs transmit motor or sensory messages (or both) primarily between the brain or brain stem and the head and neck. All CNs with the exception of the vagus (CN X) which extends through the abdominal and thoracic cavities, exit from the midbrain, pons or medulla oblongata of the brain stem. (See *Exit points for the cranial nerves*.)

Spinal nerves

With the exception of the first pair of spinal nerves (C1), which exit the spinal cord above the first cervical vertebra, each of the remaining 30 pairs of spinal nerves is generally named after the vertebrae immediately below the nerve's exit point from the spinal cord. From top to bottom they're designated as C1 through S5 and the coccygeal nerve. Each spinal nerve consists of afferent (sensory) and efferent (motor) neurones. The regions supplied by the sensory neurones are called dermatomes. (See *The spinal nerves*, page 76.)

Autonomic nervous system

The ANS *innervates* (supplies nerves to) all internal organs from the brain stem and spinal cord. Sometimes known as *visceral efferent nerves*, the nerves of the ANS carry messages to the viscera from the brain stem and neuroendocrine regulatory centres. The ANS has two major subdivisions: the *sympathetic* (thoracolumbar) nervous system and *parasympathetic* (craniosacral) nervous system.

In many instances, when one system excites, the other inhibits. Through this dual innervation, the two divisions counterbalance each other's activities to keep body systems running smoothly.

Each spinal nerve gets its name from the vertebra immediately *below* its exit point from the spinal cord.

Zoom in

Exit points for the cranial nerves

As this illustration reveals, 10 of the 12 pairs of CNs exit from the brain stem. The remaining two pairs—the olfactory and optic nerves—exit from the forebrain. Numbering of the CNs relates to their emergence from anterior to posterior.

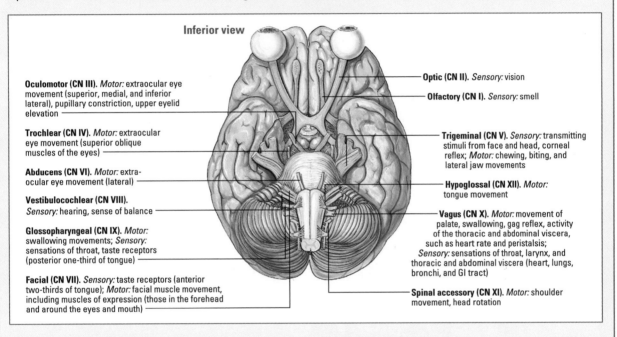

Inferior view

Oculomotor (CN III). *Motor:* extraocular eye movement (superior, medial, and inferior lateral), pupillary constriction, upper eyelid elevation

Trochlear (CN IV). *Motor:* extraocular eye movement (superior oblique muscles of the eyes)

Abducens (CN VI). *Motor:* extra-ocular eye movement (lateral)

Vestibulocochlear (CN VIII). *Sensory:* hearing, sense of balance

Glossopharyngeal (CN IX). *Motor:* swallowing movements; *Sensory:* sensations of throat, taste receptors (posterior one-third of tongue)

Facial (CN VII). *Sensory:* taste receptors (anterior two-thirds of tongue); *Motor:* facial muscle movement, including muscles of expression (those in the forehead and around the eyes and mouth)

Optic (CN II). *Sensory:* vision

Olfactory (CN I). *Sensory:* smell

Trigeminal (CN V). *Sensory:* transmitting stimuli from face and head, corneal reflex; *Motor:* chewing, biting, and lateral jaw movements

Hypoglossal (CN XII). *Motor:* tongue movement

Vagus (CN X). *Motor:* movement of palate, swallowing, gag reflex, activity of the thoracic and abdominal viscera, such as heart rate and peristalsis; *Sensory:* sensations of throat, larynx, and thoracic and abdominal viscera (heart, lungs, bronchi, and GI tract)

Spinal accessory (CN XI). *Motor:* shoulder movement, head rotation

Sympathetic nervous system

Sympathetic nerves, the *pre-ganglionic neurones* exit the spinal cord between the levels of the first thoracic and second lumbar vertebrae.

Ganglia branch out

When they leave the spinal cord, pre-ganglionic neurones enter small ganglia near the cord. The ganglia form a chain that spreads the impulse

Body shop

The spinal nerves

There are 31 pairs of spinal nerves. After leaving the spinal cord, many nerves join together to form networks called *plexuses*, as this illustration shows.

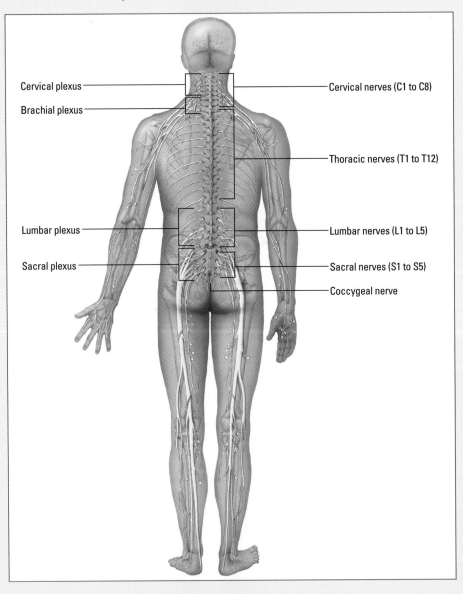

Cervical plexus

Brachial plexus

Lumbar plexus

Sacral plexus

Cervical nerves (C1 to C8)

Thoracic nerves (T1 to T12)

Lumbar nerves (L1 to L5)

Sacral nerves (S1 to S5)

Coccygeal nerve

to *postganglionic neurones*. Postganglionic neurones reach many organs and glands and can produce widespread, generalised physiological responses. These responses include:

- vasoconstriction and vasodilation
- elevated blood pressure
- enhanced blood flow to skeletal muscles
- increased heart rate and contractility
- increased respiratory rate
- smooth-muscle relaxation of the bronchioles, GI tract and urinary tract
- sphincter contraction
- pupillary dilation and ciliary muscle relaxation
- increased sweat gland secretion
- reduced pancreatic secretion.

My postganglionic neurones are making my blood pressure rise!

Parasympathetic nervous system

Fibres of the parasympathetic nervous system leave the CNS by way of the cranial nerves from the midbrain and medulla and the spinal nerves between the second and fourth sacral vertebrae (S2 to S4).

See you later, CNS

After leaving the CNS, the long preganglionic fibre of each parasympathetic nerve travels to a ganglion near a particular organ or gland. The short postganglionic fibre enters the organ or gland. This creates a more specific response involving only one organ or gland.

Such a response might be:

- reduction in heart rate
- bronchial smooth-muscle constriction
- increased GI tract tone and peristalsis, with sphincter relaxation
- increased bladder tone and urinary system sphincter relaxation
- vasodilation of external genitalia causing erection of the penis and clitoris
- pupil constriction
- increased pancreatic, salivary, lacrimal and GI tract secretions.

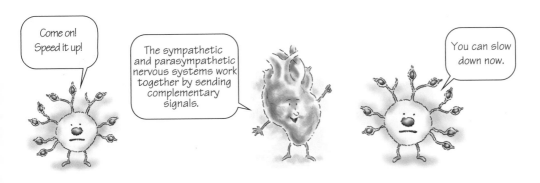

Come on! Speed it up!

The sympathetic and parasympathetic nervous systems work together by sending complementary signals.

You can slow down now.

Special sense organs

Sensory stimulation allows the body to interact with the environment. The sensory neurones act as receptors for specific stimuli, sending messages to different parts of the brain. The brain also receives stimulation from the special sense organs—the eyes, ears and gustatory and olfactory organs.

Eye

The *eyes* are the organs of vision. They contain about 70% of the body's sensory receptors. Although each eye measures about 2.5 cm in diameter, only its anterior surface is visible. Extraocular and intraocular eye structures work together for proper eye function.

Extraocular eye structures
Extraocular muscles hold the eyes in place and control their movement. Their coordinated action keeps both eyes parallel and creates binocular vision. These muscles have mutually antagonistic actions: as one muscle contracts, its opposing muscle relaxes.

Extraocular structures include the eyelids, conjunctivae and lacrimal apparatus. Together with the extraocular muscles, these structures support and protect the eyeball.

Eyelids
The *eyelids* (also called the *palpebrae*) are loose folds of skin that cover the anterior portion of the eye. The lid margins contain hair follicles, which contain eyelashes and different types of glands.

Conjunctivae
The *conjunctivae* are the thin mucous membranes that line the inner surface of each eyelid and the anterior portion of the sclera. Conjunctivae guard the eye from invasion by foreign matter. The *palpebral conjunctiva—*the portion that lines the inner surface of the eyelids—appears shiny pink or red. The *bulbar* (or ocular) *conjunctiva*, which joins the palpebral portion and covers the exposed part of the sclera, contains many small, normally visible blood vessels.

Lacrimal apparatus
The structures of the *lacrimal apparatus* (lacrimal glands, puncta, lacrimal sac and nasolacrimal duct) lubricate and protect the cornea and conjunctivae by producing and absorbing tears. Tears keep the cornea

The eyelids protect the eyes and to some extent regulate the amount of light, that can fall on the eye. They also distribute tears over the eyes.

We just left the nasolacrimal duct. Last stop, the nose.

and conjunctivae moist. Tears also contain *lysozyme*, an enzyme that provides some protection against bacterial invasion.

Cry me a river

As the eyelids blink, they direct the flow of tears from the lacrimal ducts to the *inner canthus*, the medial angle between the eyelids. Tears pool at the inner canthus and drain through the *punctum*, a tiny opening. From there they flow through the *lacrimal canals* into the lacrimal sac. Lastly, they drain through the nasolacrimal duct and into the nose. (See *A close look at tears*, page 80.)

Sclera

The white sclera is an opaque, fibrous layer that forms the major portion of the connective tissue coat of the eyeball. It is continuous with the *dura mater* and the transparent *cornea*, and helps maintain the shape of the eye, offering resistance to internal and external forces, and provides an attachment for the extraocular muscle insertions. The sclera is perforated by many nerves and vessels passing through the posterior scleral foramen, the hole that is formed by the optic nerve.

Intraocular eye structures

Intraocular structures within the eyeball are directly involved with vision. (See *Looking at intraocular structures*, page 81.)

Anterior segment

The cornea, iris, pupil, anterior chamber, aqueous humour, lens, ciliary body and posterior chamber are found in the anterior segment.

Cornea

The *cornea* is continuous with the sclera at the limbus, revealing the pupil and iris. A smooth, transparent tissue, the cornea has no blood supply but many nerve endings. It is highly sensitive to touch and is kept moist by tears.

Iris and pupil

The *iris* is a circular contractile disc that contains smooth and radial muscles. It has an opening in the centre for the *pupil*. Eye colour depends on the amount of pigment in the endothelial layers of the iris. Pupil size is controlled by involuntary dilatory and sphincter muscles in the posterior region of the iris that regulate light entry.

> Moisture plays an important role in the eye. For example, tears help to protect the cornea by keeping it moist.

Zoom in

A close look at tears

Tears begin in the lacrimal gland and drain through the nasolacrimal duct into the nose.

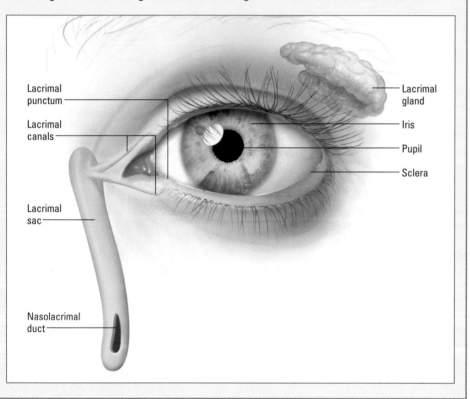

Lacrimal punctum

Lacrimal canals

Lacrimal sac

Nasolacrimal duct

Lacrimal gland

Iris

Pupil

Sclera

Anterior chamber and aqueous humour

The *anterior chamber* is a cavity bounded in front by the cornea and behind by the lens and iris. It's filled with a clear, watery fluid called *aqueous humour*.

Lens

The *lens* is situated directly behind the iris at the pupillary opening. Composed of transparent fibres in an elastic membrane called the *lens capsule*, the lens acts like a camera lens, refracting and focusing light onto the retina.

Zoom in

Looking at intraocular structures

Some intraocular structures, such as the sclera, cornea, iris, pupil and anterior chamber, are visible to the naked eye. Others, such as the retina, are visible only with an ophthalmoscope. These illustrations show the major structures within the eye.

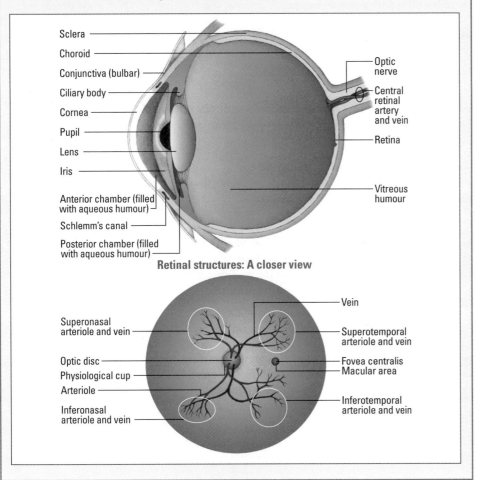

Sclera
Choroid
Conjunctiva (bulbar)
Ciliary body
Cornea
Pupil
Lens
Iris
Anterior chamber (filled with aqueous humour)
Schlemm's canal
Posterior chamber (filled with aqueous humour)

Optic nerve
Central retinal artery and vein
Retina
Vitreous humour

Retinal structures: A closer view

Superonasal arteriole and vein
Optic disc
Physiological cup
Arteriole
Inferonasal arteriole and vein

Vein
Superotemporal arteriole and vein
Fovea centralis
Macular area
Inferotemporal arteriole and vein

Want to see the sclera, cornea, iris or pupil? Just look in the mirror.

Ciliary body

The *ciliary body* (three muscles along with the iris that make up the anterior part of the *choroid*—the vascular middle layer that lies between the retina and the sclera) controls the lens thickness. Together with the coordinated action of muscles in the iris, the ciliary body regulates the light focused through the lens onto the retina.

Posterior chamber

The *posterior chamber* is a small space directly posterior to the iris but anterior to the lens. It's filled with aqueous humour.

Posterior segment

The vitreous humour, choroid and retina are found in the posterior segment.

Vitreous humour

The *vitreous humour* consists of a thick, gelatinous material that fills the space behind the lens. There, it maintains placement of the retina and the spherical shape of the eyeball.

The posterior segment of the eye has three layers: fibrous, vascular and sensory.

Choroid

The *choroid* lies beneath the sclera. It is highly vascularised providing oxygen and nutrients to the outer layers of the retina.

Retina

The *retina* is the innermost coat of the eyeball. It receives visual stimuli and sends them to the brain. The retina has a good blood supply and contains the optic disc, the physiological cup, rods and cones and the macula.

The optimal optic disc

The *optic disc* is a well-defined, 1.5 mm round or oval area on the retina. Creamy yellow to pink in colour, the optic disc allows the optic nerve to enter the retina at a point called the *nerve head*. A whitish to greyish crescent of scleral tissue may be present on the lateral side of the disc. The optic disc is lacking in rods and cones and is commonly known as the *blind spot*.

The cup within the disc

The *physiological cup* consists of a depression within the optic disc on the temporal side. It covers one-third of the centre of the disc.

Visionaries

Photoreceptor neurones called *rods* and *cones* compose the visual receptors of the retina. These receptors are responsible for vision; rods for black and white and cones for colour. The rods and cones contain specific photopigments that react to light, initiating the events that eventually lead to electrical signal generation and perception of images by the brain.

Count macula

The *macula* is lateral to the optic disc. It's slightly darker than the rest of the retina and without visible retinal vessels. A slight depression in the centre of the macula, known as the *fovea centralis*, contains the heaviest concentration of cones and is a main receptor for vision and colour.

The optic disc has no light receptors and is therefore a 'blind spot.' But I compensate so that there's usually no apparent gap in what is seen.

Vision pathway

When stimulated by light, rods and cones form impulses that are transmitted to the brain for interpretation. For correct image perception, the brain relies on structures along the vision pathway (the optic nerve, optic chiasma and retina) to create the proper visual fields.

Criss-crossing tracts

At the *optic chiasma*, fibres from the nasal aspects of both retinas cross to the opposite sides and fibres from the temporal portions remain uncrossed. These crossed and uncrossed fibres form the *optic tracts*. Injury to one of the optic nerves can cause blindness or visual impairment in the corresponding eye. An injury or lesion in the optic chiasma can cause partial vision loss (e.g. loss of the two temporal visual fields).

Focusing on the fovea centralis

Image formation begins when eye structures refract light rays from an object. Normally, the cornea, aqueous humour, lens and vitreous humour refract light rays from an object, focusing them on the fovea centralis, where an inverted and reversed image clearly forms. Within the retina, rods and cones turn the projected image into impulses and transmit them to the optic nerves.

I see! We turn light into impulses that we then transmit to the optic nerve.

Follow the tracts to the cerebral cortex

The impulses travel to the optic chiasma (where the two optic nerves unite and split again into two optic tracts) and then continue into the optic section of the cerebral cortex. There, the inverted and reversed image on the retina is processed by the brain to create an image as it truly appears in the field of vision.

Ear

The *ears* are the organs of hearing. They also maintain the body's equilibrium. The ear is divided into three main parts: external, middle and inner. (See *Ear structures*.)

Zoom in

Ear structures

The ear is the organ of hearing. The structures of its three sections—external, middle and inner—are illustrated below.

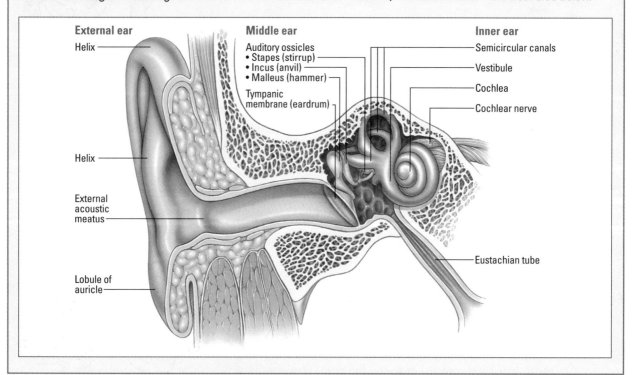

External ear

Helix

Helix

External acoustic meatus

Lobule of auricle

Middle ear

Auditory ossicles
• Stapes (stirrup)
• Incus (anvil)
• Malleus (hammer)

Tympanic membrane (eardrum)

Inner ear

Semicircular canals

Vestibule

Cochlea

Cochlear nerve

Eustachian tube

External ear structures

The *external ear* consists of the auricle and the external auditory canal. The *mastoid process* isn't part of the external ear but is an important bony landmark behind the lower part of the auricle.

Auricle

The *auricle* (pinna) is the outer, visual protrusion of the ear. It helps collect and direct incoming sound into the external auditory canal.

External auditory canal

The *external auditory canal* is a narrow chamber that connects the auricle with the tympanic membrane. This canal transmits sound waves to the tympanic membrane.

Middle ear structures

The *middle ear* is also called the *tympanic cavity*. It's an air-filled cavity within the hard portion of the temporal bone. The tympanic cavity is lined with mucosa. It's bound laterally by the tympanic membrane and medially by the oval and round windows. The Eustachian tube equalises pressure within the ear and the small bones of the middle ear conduct vibrations primarily from external sources to the inner ear structures.

Tympanic membrane

The *tympanic membrane* consists of layers of skin, fibrous tissue and a mucous membrane. It transmits sound vibrations to the middle ear.

Eustachian tube

The *Eustachian tube* extends downwards, forwards and inwards from the middle ear cavity to the nasopharynx. It has a useful function: it allows the pressure against inner and outer surfaces of the tympanic membrane to equalise, preventing rupture and allowing for proper transfer of sound waves.

Oval window

The *oval window* is an opening in the wall between the middle and inner ears into which part of the *stapes* (a tiny bone of the middle ear) fits. It transmits vibrations to the inner ear.

Round window

The *round window* is another opening in the same wall. It's enclosed by the secondary tympanic membrane. The round window helps dissipate pressure waves from the inner into the middle ear.

Small bones

The middle ear contains three small bones, called *ossicles*, which conduct vibratory motion of the tympanum to the oval window. The ossicles are:
- the *malleus* (hammer), which attaches to the tympanic membrane and transfers sound to the incus
- the *incus* (anvil), which articulates the malleus and the stapes and carries vibration to the stapes
- the *stapes* (stirrup), which connects vibratory motion from the incus to the oval window.

Inner ear structures

In the inner ear, vibration excites receptor nerve endings. A bony labyrinth and a membranous labyrinth combine to form the inner ear. The inner ear contains the vestibule, cochlea and semicircular canals.

Vestibule

The *vestibule* is located posterior to the cochlea and anterior to the semicircular canals. It serves as the entrance to the inner ear. It houses two membranous sacs: the *saccule* and *utricle*. Suspended in a fluid called *perilymph*, the saccule and utricle sense gravity changes and linear and angular acceleration.

Cochlea

The *cochlea*, a bony, spiraling cone, extends from the anterior part of the vestibule. Within it lies the *cochlear duct*, a triangular, membranous structure that houses the *organ of Corti*. The receptor organ for hearing, the organ of Corti transmits sound to the cochlear branch of the vestibulocochlear nerve (CN VIII).

Semicircular canals

The three *semicircular canals* project from the posterior aspect of the vestibule. Each canal is oriented in one of three planes: superior, posterior and lateral. The *semicircular duct* traverses the canals and connects with the utricle anteriorly. The *crista ampullaris* sits at the end of each canal and contains hair cells and support cells. It's stimulated by sudden movements or changes in the rate or direction of movement.

> For hearing to occur, sound waves travel through the ear by two pathways—air conduction and bone conduction.

Hearing pathways

For hearing to occur, sound waves travel through the ear by two pathways—air conduction and bone conduction:
- *Air conduction* occurs when sound waves travel in the air through the external and middle ear to the inner ear.
- *Bone conduction* occurs when sound waves travel through bone to the inner ear. (See *Sound transmission*.)

Interpreting the vibrations, man

Vibrations transmitted through air and bone stimulate nerve impulses in the inner ear. The cochlear branch of the vestibulocochlear nerve transmits these impulses to the cerebral temporal lobes for interpretation as sound.

Nose and mouth

The *nose* is the sense organ for smell. The mucosal epithelium that lines the uppermost portion of the nasal cavity houses receptors for fibres of the olfactory nerve (CN I).

Now I get it!

Sound transmission

The illustration below details the transmission of sound through the structures of the middle and inner ear.

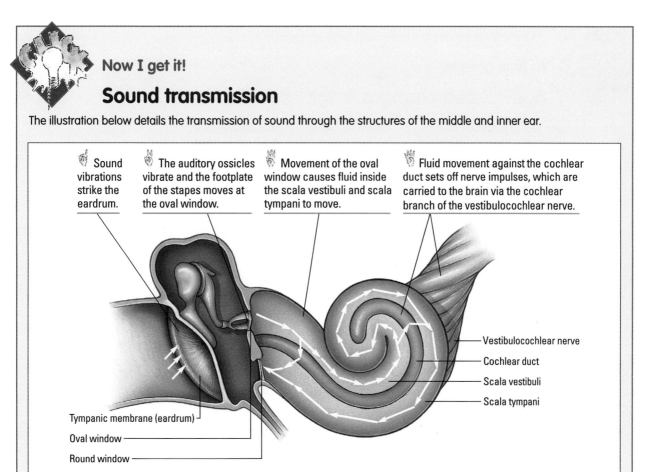

Sound vibrations strike the eardrum.

The auditory ossicles vibrate and the footplate of the stapes moves at the oval window.

Movement of the oval window causes fluid inside the scala vestibuli and scala tympani to move.

Fluid movement against the cochlear duct sets off nerve impulses, which are carried to the brain via the cochlear branch of the vestibulocochlear nerve.

Vestibulocochlear nerve

Cochlear duct

Scala vestibuli

Scala tympani

Tympanic membrane (eardrum)

Oval window

Round window

Good old olfactory

These receptors, called *olfactory* (smell) *receptors*, consist of hair cells, which are highly sensitive but easily fatigued. They're stimulated by the slightest odours but stop sensing even the strongest smells after a short time.

Slip of the tongue

The tongue and the roof of the mouth contain most of the receptors for the taste nerve fibres (located in branches of CNs VII and IX). Called *taste buds*, these receptors are stimulated by chemicals. They respond to four taste sensations: sweet, sour, bitter and salty. All the other flavours a person senses result from a combination of olfactory-receptor and taste-bud stimulation.

As time goes by . . .

Age-related changes in the special senses

With advancing years the senses become less acute, and it becomes more difficult to distinguish detail, leading to a sense of isolation.

All of the senses receive information of some type from the environment (light, sound vibrations and so on). This is converted to a nerve impulse and carried to the brain, where it is interpreted into a meaningful sensation.

Everyone requires a certain minimum amount of stimulation before a sensation is perceived. This minimum level is called the threshold. Aging increases this threshold, so the amount of sensory input needed to be aware of the sensation becomes greater. Changes in the organ related to the sensation account for most of the other sensation changes.

Hearing and vision changes are the most dramatic, but all senses can be affected by aging. Fortunately, many of the aging changes in the senses can be compensated for with equipment such as glasses and hearing aids or by minor changes in lifestyle.

Quick quiz

1.   The components of the CNS include:
   A.   the spinal cord and cranial nerves.
   B.   the brain and spinal cord.
   C.   the sympathetic and parasympathetic nervous systems.
   D.   the cranial nerves and spinal nerves.

Answer: B. The two main divisions of the nervous system are the CNS, which includes the brain and spinal cord and the peripheral nervous system, which consists of the cranial nerves, spinal nerves and ANS.

2.   The brain is protected from shock and trauma by:
   A.   bones, the meninges and CSF.
   B.   grey matter, bones and the primitive structures.
   C.   the blood-brain barrier, CSF and white matter.
   D.   axons, neurones and meninges.

Answer: A. Bones (the skull and vertebral column), the meninges and CSF protect the brain from shock and trauma.

3.   The visual receptors of the retina are comprise of:
   A.   the pupil and lens.
   B.   the optic disc and optic nerve.
   C.   vitreous humour and aqueous humour.
   D.   rods and cones.

Answer: D. Photoreceptor neurones called rods and cones comprise the visual receptors of the retina.

4. The external ear consists of:
    A. the vestibule, cochlea and semicircular ducts.
    B. the tympanic membrane, oval window and round window.
    C. the auricle and external auditory canal.
    D. the malleus, incus and stapes.

Answer: C. The auricle and external auditory canal are part of the external ear.

5. The cranial nerves transmit motor and sensory messages between the:
    A. spine and body dermatomes.
    B. brain and the head and neck.
    C. viscera and the brain.
    D. brain and skeletal muscles.

Answer: B. The 12 pairs of cranial nerves transmit motor (efferent) and sensory (afferent) messages between the brain or brain stem and the head and neck.

Scoring

☆☆☆ If you answered all five questions correctly, brilliant! You're definitely brainy.
☆☆   If you answered four questions correctly, remarkable. Your grey matter is working at lightning speed.
☆   If you answered fewer than four questions correctly, don't get nervous. Give your brain another workout by reviewing the chapter again.

Just the facts

In this chapter, you'll learn:
♦ the functions of endocrine glands
♦ hormone release and transportation in the endocrine system
♦ the role of receptors in the influence of hormones on cells.

A look at the endocrine system

The three major components of the endocrine system are:
* *glands*—specialised cell clusters or organs
* *hormones*—chemical substances secreted by glands in response to stimulation
* *receptors*—protein molecules that bind specifically with other molecules, such as hormones, to trigger specific physiologic changes in a target cell.

Along with the nervous system, the endocrine system regulates and integrates the body's metabolic activities.

Glands

The major glands of the endocrine system are:
* pituitary gland
* thyroid gland
* parathyroid glands
* adrenal glands
* pancreas
* thymus
* pineal gland
* gonads (ovaries and testes). (See *Components of the endocrine system*.)

Body Shop

Components of the endocrine system

Endocrine glands secrete hormones directly into the bloodstream to regulate body function. This illustration shows the location of the major endocrine glands.

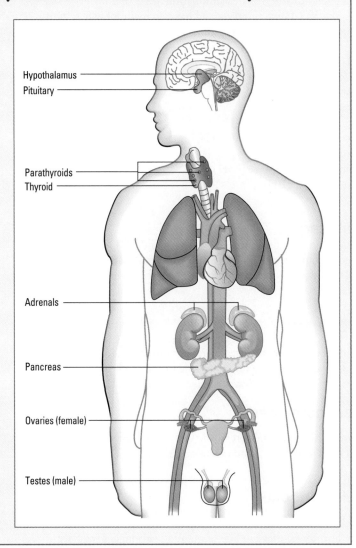

Hypothalamus
Pituitary
Parathyroids
Thyroid
Adrenals
Pancreas
Ovaries (female)
Testes (male)

The pituitary gland

The *pituitary gland* (also called the *hypophysis* or *master gland*) rests in the *sella turcica*, a depression in the sphenoid bone at the base of the brain. This pea-sized gland connects with the hypothalamus via the infundibulum stalk, from which it receives chemical and nervous stimulation. (See *How the hypothalamus affects endocrine activities*.) The pituitary gland has two main lobular regions: the anterior pituitary and the posterior pituitary.

Anterior pituitary

The *anterior pituitary* (adenohypophysis) is the larger lobe of the gland. It produces at least six hormones:
- growth hormone (GH)
- thyroid-stimulating hormone (TSH)
- adrenocorticotrophic hormone (ACTH)
- gonadotropins
  - follicle-stimulating hormone (FSH)
  - luteinising hormone (LH)
- prolactin.

Posterior pituitary

The *posterior pituitary* makes up about 25% of the gland. It does not synthesise hormones but rather serves as a storage area for antidiuretic hormone (ADH), also known as *vasopressin*, and oxytocin, which are produced by the hypothalamus.

Thyroid gland

The *thyroid* lies directly below the larynx, and lies on either side of the trachea. Its two lateral lobes—one on either side of the trachea—join with a narrow tissue bridge, called the *isthmus*, to give the gland its butterfly shape.

One gland . . . different hormones

The two lobes of the thyroid function as one unit to produce the hormones *triiodothyronine* (T_3), *thyroxine* (T_4) and *calcitonin*. (See *Thyroid stimulation*, page 94.)

T_3 and T_4 equal thyroid hormone

T_3 and T_4 are collectively referred to as the *thyroid hormones*. As the body's major metabolic hormones, these iodine-containing substances regulate metabolism by primarily speeding up cellular respiration. They are also necessary for normal development of the nervous and musculoskeletal system.

Now I get it!

How the hypothalamus affects endocrine activities

Hypothalamus and pituitary

Anterior and posterior pituitary secretions are controlled by signals from the hypothalamus:

- As shown on the left side of the illustration below, different hypothalamic neurones produce antidiuretic hormone (ADH) and oxytocin, which travel down the axons and are stored in secretory granules in nerve endings in the posterior pituitary for later release.
- As shown on the right side of the illustration below, the hypothalamus stimulates the anterior pituitary's production of its many hormones. A hypothalamic neurone manufactures inhibitory and stimulatory hormones and secretes them into a capillary of the portal system. The hormones travel down the pituitary stalk to the anterior pituitary. There, they cause inhibition or release of many pituitary hormones, including corticotropin, thyroid-stimulating hormone, growth hormone, follicle-stimulating hormone, luteinising hormone and prolactin.

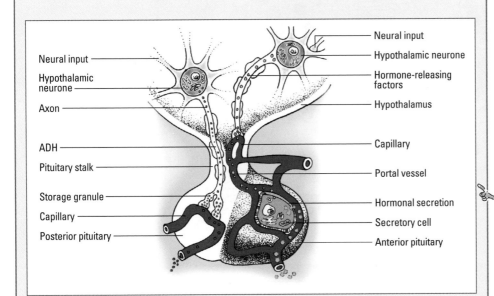

Neural input — Hypothalamic neurone — Axon — ADH — Pituitary stalk — Storage granule — Capillary — Posterior pituitary

Neural input — Hypothalamic neurone — Hormone-releasing factors — Hypothalamus — Capillary — Portal vessel — Hormonal secretion — Secretory cell — Anterior pituitary

The hypothalamus is the integrative centre for the endocrine and autonomic nervous systems.

Now I get it!

Thyroid stimulation

Thyroid cells store a hormone precursor, thyroglobulin, which contains iodine and is stored in the colloid of the thyroid gland. When stimulated by thyroid-stimulating hormone (TSH), a follicular cell (shown below) takes up some stored thyroglobulin by *endocytosis*—the reverse of exocytosis.

The cell membrane extends fingerlike projections into the colloid, and then pulls portions of it back into the cell. Lysosomes fuse with the colloid, which is then degraded into triiodothyronine (T_3) and thyroxine (T_4). These thyroid hormones are released into the circulation and lymphatic system by exocytosis.

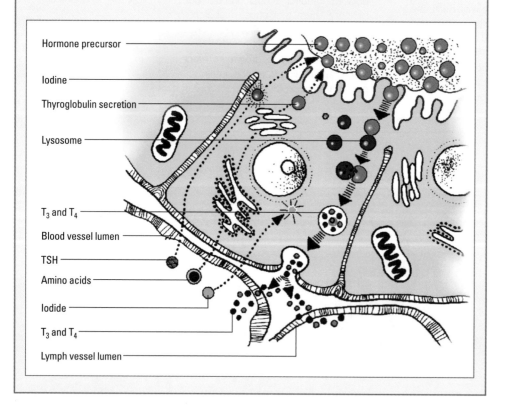

Calcium balancing act

Calcitonin assists in the maintenance of blood calcium levels. It does this by inhibiting the release of calcium from bone—opposite to the effects of the hormones of the parathyroid gland. Secretion of calcitonin is controlled by the calcium concentration of the fluid surrounding the thyroid cells.

Parathyroid glands

The *parathyroid glands* are the body's smallest known endocrine glands. These glands are embedded on the posterior surface of the thyroid, usually one in each corner.

Calcium cohort

Working together as a single gland, the parathyroid glands produce *parathyroid hormone* (PTH). The main function of PTH is to help regulate the blood's calcium balance. PTH raises blood calcium by reabsorbing calcium from the kidneys and gastrointestinal tract. PTH will also break down bone to restore calcium levels in the blood. The urinary loss of calcium and magnesium ions is controlled by PTH, as is the loss of phosphate.

The main function of PTH is to help regulate the blood's calcium balance.

Adrenal glands

The two *adrenal glands* each lie on top of a kidney. These glands contain two distinct structures—the adrenal cortex and the adrenal medulla—that function as separate endocrine glands.

Adrenal cortex

The *adrenal cortex* is the large outer layer. It forms the bulk of the adrenal gland. It has three zones, or cell layers:

The *zona glomerulosa*, the outermost zone, produces mineralocorticoids (primarily aldosterone) that help maintain fluid balance by increasing sodium reabsorption excretion. This ultimately affects the overall management of blood volume and pressure.

The *zona fasciculata*, the middle and largest zone, produces the glucocorticoids cortisol, cortisone and corticosterone as well as small amounts of the sex hormones androgen and oestrogen. Glucocorticoids are primarily involved in the regulation of metabolism and resistance to stress. However, they also have anti-inflammatory effects and can depress immune function.

The *zona reticularis*, the innermost zone, produces some sex hormones.

Adrenal medulla

The *adrenal medulla*, or inner layer of the adrenal gland, functions as part of the sympathetic nervous system and produces two catecholamines: adrenaline (epinephrine) and noradrenaline (norepinephrine). Because catecholamines play an important role in the autonomic nervous system (ANS), the adrenal medulla is considered a neuroendocrine structure.

Memory jogger

To remember the location of the adrenal glands, think **Add-Renal.** They're 'added' to the renal organs, the kidneys.

Pancreas

The *pancreas*, a triangular organ, is nestled in the curve of the duodenum, stretching horizontally behind the stomach and extending to the spleen.

Endo and exo

The pancreas performs both endocrine and exocrine functions. As its endocrine function, the pancreas secretes hormones, while its exocrine function is secreting digestive enzymes. *Acinar cells* make up most of the gland and help regulate pancreatic exocrine function. (See *Pancreas stimulation*.)

Islands in an acinar sea

The endocrine cells of the pancreas are called the *islet cells*, or *islets of Langerhans*. These cells exist in clusters and are found scattered among the acinar cells. The islets contain alpha, beta and delta cells that produce important hormones:

- Alpha cells produce *glucagon*, a hormone that raises the blood glucose level by triggering the breakdown of glycogen to glucose.
- Beta cells produce *insulin*. Insulin lowers the blood glucose level by stimulating the conversion of glucose to glycogen.
- Delta cells produce *somatostatin*. Somatostatin inhibits the release of GH, ACTH, insulin and glucagon.

Thymus

The *thymus* is located posterior to the sternum and contains lymphatic tissue. It reaches maximal size at puberty and then starts to atrophy (degenerate).

Developing Mr. T cells, fool!

Since the thymus is the site for T cell development (which are important in cell-mediated immunity), the thymus has been proposed to have a major involvement in immune function. However, the thymus also produces several hormones including thymosin and thymopoietin. These hormones promote growth of peripheral lymphoid tissue.

Pineal gland

The tiny *pineal gland* lies at the back of the third ventricle of the brain. It produces the hormone *melatonin*, primarily during the dark hours of the day. Little hormone is produced during daytime hours, thereby influencing sleep-wake cycles. Melatonin is thought to regulate circadian rhythms, body temperature, cardiovascular function and reproduction.

Gonads

The *gonads* include the ovaries (in females) and the testes (in males).

Ovaries

The *ovaries* are paired, oval glands that are situated on either side of the uterus. They produce ova (eggs) and the steroidal hormones oestrogen and progesterone. These hormones have four functions:

☝ They promote development and maintenance of female sex characteristics.

✌ They regulate the menstrual cycle.

🖖 They maintain the uterus for pregnancy.

🖐 Along with other hormones, they prepare the mammary glands for lactation.

Testes

The *testes* are paired structures that lie in an extra-abdominal pouch (scrotum) in the male. They produce spermatozoa and the male sex hormone testosterone. Testosterone stimulates and maintains masculine sex characteristics and triggers the male sex drive.

Hormones

Hormones are complex chemical substances that trigger or regulate the activity of an organ or a group of cells. Hormones are classified by their molecular structure as *polypeptides*, *steroids*, *amines* and *eicosanoids*.

Polypeptides

Polypeptides are protein compounds made of many amino acids that are connected by peptide bonds. They include:
- anterior pituitary hormones (GH, TSH, ACTH, FSH, LH and prolactin)
- posterior pituitary hormones (ADH and oxytocin)
- parathyroid hormone (PTH)
- pancreatic hormones (insulin and glucagon).

Steroids

Steroids are derived from cholesterol. They include:
• adrenocortical hormones secreted by the adrenal cortex (aldosterone and cortisol)
• sex hormones secreted by the gonads (oestrogen and progesterone in females and testosterone in males).

Amines

Amines are derived from *tyrosine*, an essential amino acid found in most proteins. They include:
• thyroid hormones (T_4 and T_3)
• the catecholamines adrenaline (epinephrine), noradrenaline (norepinephrine) and dopamine.

Eicosanoids

The *Eicosanoids* are a diverse group of molecules derived from arachadonic acid which have a wide range of biological effects and are mediators of inflammation and pain impulse transmission. They include:
• prostaglandins
• leukotrienes

Hormone release and transport

Although all hormone release results from endocrine gland stimulation, release patterns of hormones vary greatly. For example:
• ACTH (secreted by the anterior pituitary) and cortisol (secreted by the adrenal cortex) are released in spurts in response to body rhythm cycles. Levels of these hormones peak in the early morning and are lowest in the evening.
• Secretion of PTH (by the parathyroid gland) and prolactin (by the anterior pituitary) occurs fairly evenly throughout the day. However, levels of prolactin also change in response to oestrogen levels in females.
• Secretion of insulin by the pancreas can occur at a steady rate or sporadically, depending on blood glucose levels.

Thyroid and steroid hormones circulate while bound to plasma proteins, whereas catecholamines and most polypeptides aren't protein-bound.

Hormonal action

When a hormone reaches its target site, it binds to a specific receptor on the cell membrane or within the cell. Polypeptides and some amines bind to membrane receptor sites. The smaller, more lipid-soluble steroids and thyroid hormones diffuse through the cell membrane and bind to intracellular receptors.

Right on target!

After binding occurs, each hormone produces unique physiologic changes, depending on its target site and its specific action at that site. A particular hormone may have different effects at different target sites.

Hormonal regulation

To maintain the body's delicate equilibrium, a feedback mechanism regulates hormone production and secretion. The mechanism involves hormones, blood chemicals and metabolites, and the nervous system. This system may be simple or complex. (See *The feedback loop*, page 100.)

Signalling secretory cells

Cells producing hormones may be near or at a distance from their target tissue. When cells receive a stimulus (which may be neural or hormonal), the cell releases its hormones into extracellular fluid. The hormone then travels to the target tissue where it binds to its receptors and causes a physiological response which in turn can inhibit the cell from releasing more hormone through negative feedback. (See *A close look at target cells*, page 101.)

Mechanisms that control hormone release

Four basic mechanisms control hormone release:

 the pituitary-target gland axis

 the hypothalamic-pituitary-target gland axis

 chemical regulation

 nervous system regulation.

Pituitary-target gland axis

The pituitary gland regulates other endocrine glands—and their hormones—through secretion of *trophic hormones* (releasing and inhibiting hormones). These hormones include:

- corticotropin, which regulates adrenocortical hormones
- TSH, which regulates T_4 and T_3
- LH, which regulates gonadal hormones.

Now I get it!

The feedback loop

This diagram shows the negative feedback mechanism that helps regulate the endocrine system. Negative feedback mechanisms, irrespective of their level of complexity, all comprise of the same essential elements—a detector, control centre and an effector.

Types of feedback

Simple feedback occurs when the level of one substance regulates the secretion of hormones (simple loop). For example, a low serum calcium level stimulates the parathyroid gland to release parathyroid hormone (PTH). PTH, in turn, promotes resorption of calcium from the GI tract, kidneys and bones. A high serum calcium level inhibits PTH secretion.

Hypothalamic stimulation can trigger a *complex feedback* mechanism. First, the hypothalamus sends releasing and inhibiting factors or hormones to the anterior pituitary. In response, the anterior pituitary secretes tropic hormones, such as growth hormone (GH), prolactin (PRL), corticotropin, thyroid-stimulating hormone (TSH), follicle-stimulating hormone (FSH) and luteinising hormone (LH). At the appropriate target gland, these hormones stimulate the target organ to release other hormones that regulate various body functions. When these hormones reach normal levels in body tissue, a feedback mechanism inhibits further hypothalamic and pituitary secretion.

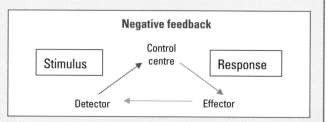

Negative feedback

Stimulus → Control centre → Response

Detector ← Effector

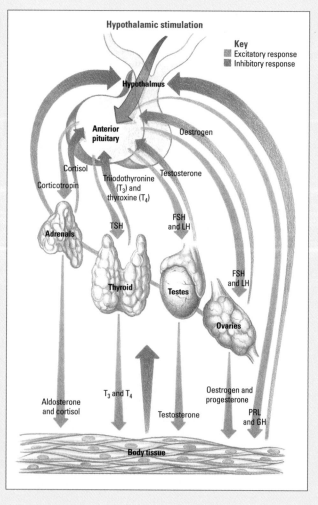

Now I get it!

A close look at target cells

A hormone acts only on cells that have receptors specific to that hormone. The sensitivity of a target cell depends on how many receptors it has for a particular hormone. The more receptor sites, the more sensitive the target cell.

A hormone acts only on a cell that has a receptor specific to that hormone.

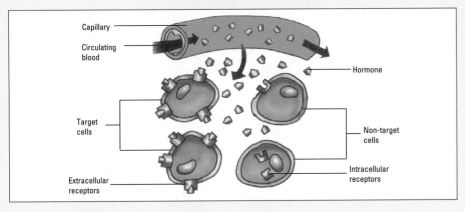

Capillary

Circulating blood

Hormone

Target cells

Non-target cells

Intracellular receptors

Extracellular receptors

Picking up feedback

The pituitary gland gets feedback about target glands by continuously monitoring levels of hormones produced by these glands (or the physiological effects the hormones). If a change occurs, the pituitary gland corrects it in one of two ways:
• by increasing the trophic hormones, which stimulate the target gland to increase production of some target gland hormones
• by decreasing the trophic hormones, thereby decreasing target gland stimulation and target gland hormone levels.

Hypothalamic-pituitary-target gland axis

The hypothalamus also produces trophic hormones that regulate anterior pituitary hormones. By controlling anterior pituitary hormones, which regulate the target gland hormones, the hypothalamus affects target glands as well.

Chemical regulation

Endocrine glands not controlled by the pituitary gland may be controlled by specific substances that trigger gland secretions. For example, blood glucose level is a major regulator of glucagon and insulin release. When blood glucose level rises, the pancreas is stimulated to increase insulin secretion and suppress glucagon secretion. A depressed level of blood glucose, on

It says here that specific substances, such as blood glucose, can trigger gland secretions.

As time goes by . . .

Endocrine changes with aging

As a person ages, normal changes in endocrine function include reduced progesterone production, a 50% decline in serum aldosterone levels and a 25% decrease in cortisol secretion rate.

Another common and important endocrine change in elderly people is a change in glucose metabolism in response to stress. Normally, both young adults and elderly people have similar fasting blood glucose levels. However, under stressful conditions, an elderly person's blood glucose level rises higher and remains elevated longer than does a younger adult's.

In females, the onset of the menopause and reduced oestrogen production promotes osteoporosis—a general reduction in bone density.

the other hand, triggers increased glucagon secretion and suppresses insulin secretion. (See *Endocrine changes with aging*.)

Nervous system regulation

The central nervous system (CNS) helps to regulate hormone secretion in several ways.

Hypothalamus has control . . .

The hormones synthesised by the neurosecretory cells of the hypothalamus pass in their axons to the posterior pituitary where they are stored in the axonal terminals. Because hypothalamic nerve cells stimulate the posterior pituitary to release ADH and oxytocin, these hormones are ultimately controlled by the CNS.

. . . but stimuli matter, too

Nervous system stimuli—such as hypoxia (oxygen deficiency), nausea, pain, stress and certain drugs—also affect ADH levels.

ANS steers this ship . . .

The ANS controls catecholamine secretion by the adrenal medulla.

. . . while stress spikes ACTH

The nervous system also affects other endocrine hormones. For example, stress, which leads to sympathetic stimulation, causes the pituitary to release ACTH.

Quick quiz

1. The purpose of the endocrine system is to:
    A. deliver nutrients to the body's cells.
    B. regulate and integrate the body's metabolic activities.
    C. eliminate waste products from the body.
    D. control the body's temperature and produce blood cells.

Answer: B. Along with the nervous system, the endocrine system regulates and integrates the body's metabolic activities.

2. The mechanism that most commonly regulates the endocrine system is called the:
    A. transport mechanism.
    B. self-regulation mechanism.
    C. feedback mechanism.
    D. pituitary-target gland axis.

Answer: C. The negative feedback mechanism helps regulate the endocrine system by signalling to the endocrine glands the need for changes in hormone levels.

3. The gland that produces glucagon is the:
    A. pancreas.
    B. thymus.
    C. adrenal gland.
    D. pituitary gland.

Answer: A. The alpha cells of the pancreas produce glucagon, a hormone that raises the blood glucose level by triggering the breakdown of glycogen to glucose.

4. Pituitary hormones are controlled by the:
    A. pancreas.
    B. hypothalamus.
    C. thyroid gland.
    D. parathyroid glands.

Answer: B. The hypothalamus controls pituitary hormones.

Scoring

☆☆☆   If you answered all four questions correctly, astonishing! Your brain cells must be on steroids!

☆☆   If you answered three questions correctly, bon voyage! You've just won a trip to the islets of Langerhans!

☆   If you answered fewer than three questions correctly, don't moan over these hormones. A quick revision session and you'll be back in control!

Cardiovascular system 6

Just the facts

In this chapter, you'll learn:
♦ structures of the heart and their functions
♦ the heart's conduction system
♦ the flow of blood through the heart and the body.

A look at the cardiovascular system

The cardiovascular system (sometimes called the *circulatory system*) primarily consists of the *heart, blood vessels* and *lymphatics*. This network brings life-sustaining oxygen and nutrients to the body's cells, removes metabolic waste products and carries hormones from one part of the body to another.

Doing double duty

The heart is actually two separate pumps: The right side pumps the blood to the lungs to receive oxygen and the left side pumps the oxygenated blood to the rest of the body.

Where the heart lies

About the size of a closed fist, the heart lies beneath the sternum in the *mediastinum* (the cavity between the lungs), between the second and sixth ribs. In most people, the heart rests obliquely, with its right side more anterior than the left. Because of its oblique angle, the heart's broad part or top is at its upper right, and its pointed end (apex) is at its lower left. The apex is the *point of maximal impulse*, where the heartbeat can be most readily felt as it pushes against the chest wall with each beat.

I usually lie slightly to the left of the chest midline ... but this will do for now!

Heart structure

Surrounded by a sac called the *pericardium*, the heart has a wall made up of three layers: the *myocardium*, *endocardium* and *epicardium*. Within the heart lie four chambers (two atria and two ventricles) and four valves (two atrioventricular [AV] and two semilunar valves). (See *Inside the heart*, page 106.)

Pericardium

The *pericardium* is a fibroserous sac that surrounds the heart and the roots of the *great vessels* (those vessels that enter and leave the heart). It consists of the fibrous pericardium and the serous pericardium.

Fibrous fits freely

The *fibrous pericardium*, composed of tough, white fibrous tissue, fits loosely around the heart, protecting it.

Serous is smooth

The *serous pericardium*, the thin, smooth inner portion, has two layers:
- The *parietal layer* lines the inside of the fibrous pericardium.
- The *visceral layer* adheres to the surface of the heart.

The space between

Between the fibrous and serous pericardium is the *pericardial space*. This space contains *pericardial fluid* that lubricates the surfaces of the space and allows the heart to move easily during contraction.

I'm supported and protected by a tough, fibrous sac, but I stay comfortable because it has a smooth inner lining and lubricating fluid.

The wall

The wall of the heart consists of three layers:

The *epicardium*, the outer layer (and the visceral layer of the serous pericardium), is made up of squamous epithelial cells overlying connective tissue.

The *myocardium*, the middle layer, forms most of the heart wall. It has striated muscle fibres that cause the heart to contract.

The *endocardium*, the heart's inner layer, consists of endothelial tissue with small blood vessels and bundles of smooth muscle.

Zoom in

Inside the heart

Within the heart lie four chambers (two atria and two ventricles) and four valves (two atrioventricular and two semilunar valves). A system of blood vessels carries blood to and from the heart.

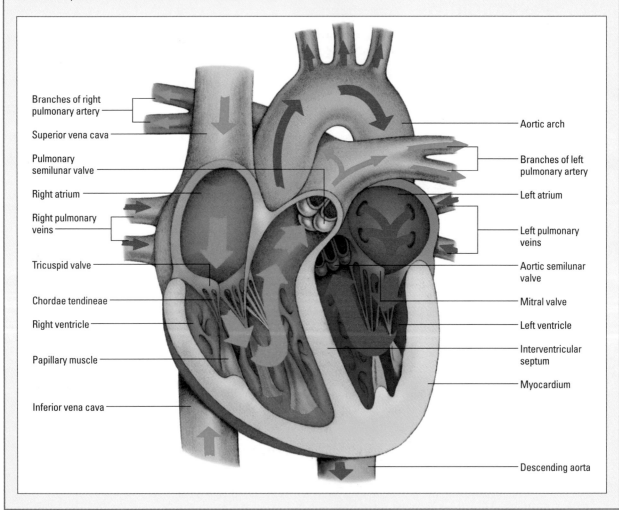

Branches of right pulmonary artery

Superior vena cava

Pulmonary semilunar valve

Right atrium

Right pulmonary veins

Tricuspid valve

Chordae tendineae

Right ventricle

Papillary muscle

Inferior vena cava

Aortic arch

Branches of left pulmonary artery

Left atrium

Left pulmonary veins

Aortic semilunar valve

Mitral valve

Left ventricle

Interventricular septum

Myocardium

Descending aorta

The chambers

The heart contains four hollow chambers: two atria (singular: atrium) and two ventricles.

Upstairs . . .

The *atria*, the upper chambers, are separated by the *interatrial septum*. They receive blood returning to the heart and supply blood to the ventricles.

. . . where the blood comes in

The *right atrium* receives blood from the *superior* and *inferior venae cavae*. The *left atrium*, which is smaller but has thicker walls than the right atrium, forms the uppermost part of the heart's left border. It receives blood from the four pulmonary veins.

Downstairs . . .

The *right* and *left ventricles*, separated by the *interventricular septum*, make up the two lower chambers. The ventricles receive blood from the atria. Composed of highly developed musculature, the ventricles are larger and have thicker walls than the atria.

. . . where the blood goes out

The right ventricle pumps blood to the lungs. The left ventricle, which has a much thicker wall than the right, pumps blood into the aorta.

The valves

The heart contains four valves: two *AV valves* and two crescent-shaped *semilunar valves*.

One way only

The valves allow forward flow of blood through the heart and prevent backward flow. They open and close in response to pressure changes caused by ventricular contraction and blood ejection.

The two AV valves separate the atria from the ventricles. The right AV valve, called the *tricuspid valve*, prevents backflow from the right ventricle into the right atrium. The left AV valve, called the *bicuspid* or *mitral valve*, prevents backflow from the left ventricle into the left atrium.

One of the two semilunar valves is the *pulmonary valve*, which prevents backflow from the pulmonary arteries into the right ventricle. The other semilunar valve is the *aortic valve*, which prevents backflow from the aorta into the left ventricle.

Memory jogger

If you can remember that there are two distinct heart sounds, you can recall that there are two sets of heart valves. Closure of the atrioventricular valves makes the first heart sound, the **lub**; closure of the semilunar valves makes the second heart sound, the **dub.**

On the cusps

The tricuspid valve has three triangular *cusps*, or leaflets. The mitral valve, also called the *bicuspid valve*, contains two cusps, a large anterior and a smaller posterior. *Chordae tendineae* attach the cusps of the AV valves to papillary muscles in the ventricles. The semilunar valves have three cusps that are shaped like half-moons.

Conduction system

Contraction of the heart, occurring as a result of its *conduction system*, causes blood to move throughout the body. (See *Cardiac conduction system*.)

Cardiac conduction system

Specialised fibres propagate electrical impulses throughout the heart's cells, causing the heart to contract. This illustration shows the elements of the cardiac conduction system.

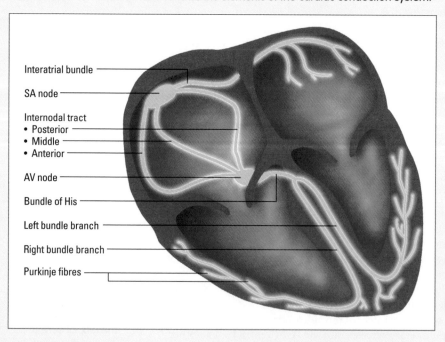

Interatrial bundle

SA node

Internodal tract
• Posterior
• Middle
• Anterior

AV node

Bundle of His

Left bundle branch

Right bundle branch

Purkinje fibres

Electrical impulses help me conduct myself at a special rhythm.

Setting the pace

The conduction system of the heart contains *pacemaker cells*, which have two unique characteristics:
- *automaticity*, the ability to generate an electrical impulse automatically without a nerve impulse
- *conductivity*, the ability to pass the impulse to the next cell

These actions ultimately result in the contraction of cardiac myocytes after they receive the impulse.

Feeling impulsive

The *sinoatrial (SA) node*, located in the right atrium, near the superior vena cava, is the normal pacemaker of the heart, generating an impulse between 60 and 100 times per minute. The firing of the SA node spreads an impulse throughout the right and left atria, resulting in *atrial contraction*. The electrical activity of the heart can be measured by means of an electrocardiograph. By examining the pattern and characteristics of this electrical activity, the state of the myocardium and conduction system can be determined. (*See Electrical changes in the heart*, page 110.)

Fill 'er up

The *AV node*, situated low in the *septal wall* of the right atrium, slows impulse conduction between the atria and ventricles. This 'resistor' node allows time for the contracting atria to fill the ventricles with blood before the lower chambers contract.

Time to contract

From the AV node, the impulse travels to the *intraventricular bundle* (also known as the *bundle of His*), branching off to the right and left bundles. Finally, the impulse travels to the *Purkinje fibres*, the distal portions of the left and right bundle branches. These fibres fan across the surface of the ventricles from the endocardium to the myocardium. As the impulse spreads, it causes the blood-filled ventricles to contract.

Foolproof

The conduction system has two built-in safety mechanisms. If the SA node fails to fire, the AV node will generate an impulse between 40 and 60 times per minute. If the SA node and AV node fail, the ventricles can generate their own impulse between 20 and 40 times per minute, although this is not enough to maintain cardiac output.

My conduction system has two backup impulse generators.

Now I get it!

Electrical changes in the heart

The hearts electrical activity can be measured using an electrocardiograph (ECG). Since body fluids and tissues are good conductors of electrical impulses, the attachment of electrodes to the body surface enables changes in electrical activity to be picked up and displayed as an electrocardiogram.

A normal ECG shows 5 classic waves (P, Q, R, S and T). The P wave indicates activity arising in the SA node and sweeping over the atria. The QRS complex relates to the spread of the impulse from the AV node through the AV bundle and Purkinje fibres, as well as the electrical activity of the cells that form the ventricles. The T wave indicates relaxation of the ventricular muscle.

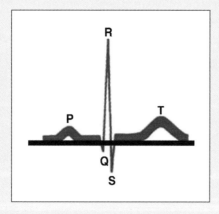

Cardiac cycle

The *cardiac cycle* is the period from the beginning of one heartbeat to the beginning of the next. During this cycle, electrical and mechanical events must occur in the proper sequence and to the proper degree to provide adequate cardiac output to the body. The cardiac cycle has two phases: *systole* and *diastole*. (See *Events in the cardiac cycle*.)

Contract . . .

At the beginning of *systole*, the ventricles contract. Increasing blood pressure in the ventricles forces the AV valves (mitral and tricuspid) to close and the semilunar valves (pulmonary and aortic) to open.

As the ventricles contract, ventricular blood pressure builds until it exceeds the pressure in the pulmonary artery and the aorta. Then the semilunar valves open, and the ventricles eject blood into the aorta and the pulmonary artery.

Now I get it!

Events in the cardiac cycle

The cardiac cycle consists of the following five events.

Isovolumetric ventricular contraction—In response to ventricular contraction, tension in the ventricles increases. This rise in pressure within the ventricles leads to closure of the atrioventricular valves (mitral and tricuspid). The pulmonary and aortic valves stay closed during the entire phase.

Ventricular ejection—When ventricular pressure exceeds aortic and pulmonary arterial pressure, the aortic and pulmonary valves open and the ventricles eject blood.

Isovolumetric relaxation—When ventricular pressure falls below the pressure in the aorta and pulmonary artery, the aortic and pulmonary valves close. All valves are closed during this phase. Atrial diastole occurs as blood fills the atria.

Atrial systole—Known as the atrial kick, atrial systole (coinciding with late ventricular diastole) supplies the ventricles with the remaining 30% of the blood for each heartbeat.

Ventricular filling—Atrial pressure exceeds ventricular pressure, which causes the left and right atrioventricular valves (mitral and tricuspid) to open. Blood then flows passively into the ventricles. About 70% of ventricular filling takes place during this phase.

. . . and release

When the ventricles empty and relax, ventricular pressure falls below the pressure in the pulmonary artery and the aorta. At the beginning of *diastole*, the semilunar valves close to prevent the backflow of blood into the ventricles, and the mitral and tricuspid valves open, allowing blood to flow into the ventricles from the atria.

When the ventricles are around 70% full, near the end of this phase, the atria contract to send the remaining blood to the ventricles. A new cardiac cycle begins as the heart enters systole again.

Cardiac output

Cardiac output refers to the amount of blood the heart pumps in 1 minute. It's equal to the heart rate multiplied by the *stroke volume*, the amount of blood ejected with each heartbeat. Stroke volume, in turn, depends on three major factors: *preload*, *contractility* and *afterload*. (See *Understanding preload, contractility and afterload*.)

A typical heart rate of 70 bpm and a stroke volume of 70 ml means I pump 4900 ml of blood around the body each minute . . . and that's when I'm resting!

Blood flow

As blood makes its way through the vascular system, it travels through three distinct types of blood vessels, involving three areas of circulation (pulmonary, cardiac and systemic).

Blood vessels

The three types of blood vessels are arteries, capillaries and veins. The structure of each type of vessel differs according to its function in the cardiovascular system and the pressure exerted by the volume of blood at various sites within the system.

Through thick . . .

Arteries have thick, muscular walls to accommodate the flow of blood at high speeds and pressures. *Arterioles* branch from the arteries and their walls also contain significant amounts of smooth muscle. They constrict or dilate to control blood flow to the *capillaries*, which have walls composed of only a single layer of endothelial cells.

. . . and thin

Venules gather blood from the capillaries; their walls contain less smooth muscle than those of arterioles. *Veins* have thinner walls than arteries but typically have larger diameters because of the low blood pressures of venous return to the heart. Also, because of the relatively low pressure of the blood that veins conduct, valves are present to prevent backflow of blood.

The autonomic nervous system increases or decreases my heart activity to meet the metabolic needs of my body. Whew!

Now I get it!

Understanding preload, contractility and afterload

If you think of the heart as a balloon, it will help you understand stroke volume.

Blowing up the balloon

Preload is the stretching of muscle fibres in the ventricles. This stretching results from blood volume in the ventricles at end-diastole. According to the *Frank-Starling law,* the more the heart muscles stretch during diastole, the more forcefully they contract during systole (and hence the greater the stroke volume). Think of preload as the balloon stretching as air is blown into it. The more air, the greater the stretch.

The balloon's stretch

Contractility refers to the inherent ability of the myocardium to contract normally. Contractility is influenced by preload. The greater the stretch, the more forceful the contraction—or, the more air in the balloon, the greater the stretch, and the farther the balloon will fly when air is allowed to expel.

The knot that ties the balloon

Afterload refers to the pressure that the ventricular muscles must generate to overcome the higher pressure in the aorta and pulmonary arteries to get the blood out of the heart. An increase in afterload acts to decrease stroke volume. *Resistance* is the knot on the end of the balloon, which the balloon has to work against to get the air out.

Taking the long way home

About 60,000 miles of arteries, arterioles, capillaries, venules and veins keep blood circulating to and from every functioning cell in the body. (See *Major blood vessels,* page 114.)

Body shop

Major blood vessels

This illustration shows the body's major arteries and veins.

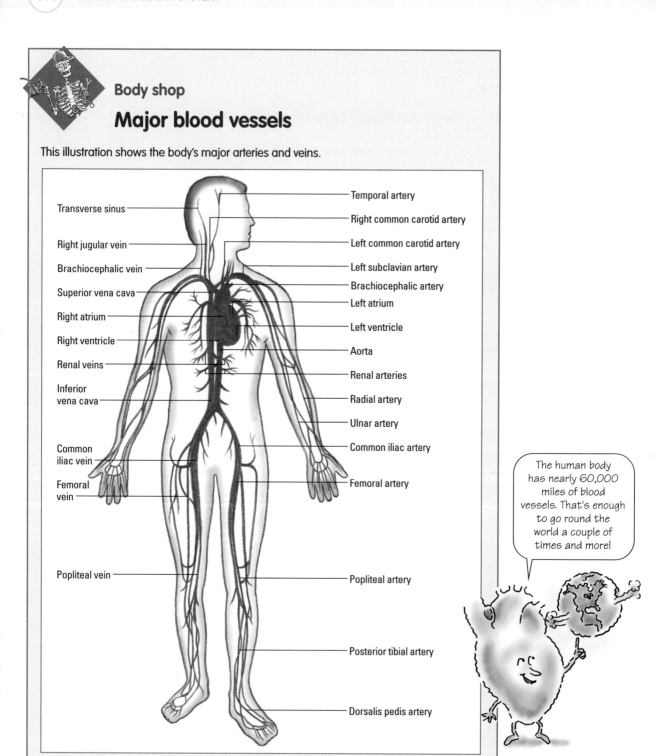

Transverse sinus

Right jugular vein

Brachiocephalic vein

Superior vena cava

Right atrium

Right ventricle

Renal veins

Inferior
vena cava

Common
iliac vein

Femoral
vein

Popliteal vein

Temporal artery

Right common carotid artery

Left common carotid artery

Left subclavian artery

Brachiocephalic artery

Left atrium

Left ventricle

Aorta

Renal arteries

Radial artery

Ulnar artery

Common iliac artery

Femoral artery

Popliteal artery

Posterior tibial artery

Dorsalis pedis artery

The human body has nearly 60,000 miles of blood vessels. That's enough to go round the world a couple of times and more!

Circulation

There are three methods of circulation that carry blood throughout the body: *pulmonary*, *systemic* and *coronary*.

Pulmonary circulation

Blood travels to the lungs to pick up oxygen and release carbon dioxide.

Returns and exchanges

As the blood moves from the heart, to the lungs and back again, it proceeds as follows:
• Deoxygenated blood travels from the right ventricle through the pulmonary valve into the *pulmonary arteries*.
• Blood passes through progressively smaller arteries and arterioles into the capillaries of the lungs.
• Blood reaches the capillaries surrounding the *alveoli* and exchanges carbon dioxide for oxygen.
• Oxygenated blood then returns via venules and veins to the *pulmonary veins*, which carry it back to the heart's left atrium.

At rest, only about 20% of my blood goes to skeletal muscles. When I exercise, that percentage can increase to 70%.

Systemic circulation

Blood pumped from the left ventricle carries oxygen and other nutrients to body cells and transports waste products for excretion.

Branching out

The major artery, the *aorta*, branches into vessels that supply specific organs and areas of the body. As it arches out of the top of the heart and down to the abdomen, three arteries branch off the top of the arch to supply the upper body with blood:
• The *left common carotid artery* supplies blood to the brain.
• The *left subclavian artery* supplies the arms.
• The *brachiocephalic artery* supplies the upper chest.
As the aorta descends through the thorax and abdomen, its branches supply the organs of the GI and genitourinary systems, spinal column and lower chest and abdominal muscles. Then the aorta divides into the *iliac arteries*, which further divide into *femoral arteries* which supply the lower limbs.

Division = addition = perfusion

As the arteries divide into smaller units, the number of vessels increases dramatically, thereby increasing the area of tissue to which blood flows, also called the *area of perfusion*.

Dilation is another part of the equation

At the end of the arterioles and the beginning of the capillaries, *sphincters* control blood flow into the tissues. These sphincters dilate to permit more flow when needed, close to shunt blood to other areas or constrict to reduce blood flow.

A large area of low pressure

Although the *capillary bed* contains the smallest vessels, it supplies blood to the largest number of cells. Capillary pressure is extremely low to allow for the exchange of nutrients, oxygen and carbon dioxide with body cells. From the capillaries, blood flows into venules and, eventually, into veins.

No backflow

Valves in the veins prevent blood backflow. Pooled blood in each valved segment is moved towards the heart by pressure from the moving volume of blood from below. The veins merge until they form two main branches, the *superior vena cava* and *inferior vena cava*, which return blood to the right atrium.

Coronary circulation

The heart relies on the coronary arteries and their branches for its supply of oxygenated blood and depends on the cardiac veins to remove deoxygenated blood. (See *Vessels that supply the heart*.)

The heart gets its share

During systole, blood is ejected into the aorta from the left ventricle. During diastole, blood flows out of the heart into the aorta and coronary arteries.

From the right . . .

The *right coronary artery* (which branches to form the *posterior interventricular artery* and the *marginal artery*) supplies blood to the right atrium, part of the left atrium, most of the right ventricle and the inferior part of the left ventricle.

. . . and from the left

The *left coronary artery*, which splits into the *anterior descending artery* and *circumflex artery*, supplies blood to the left atrium, most of the left ventricle, and most of the interventricular septum.

Like any other muscle, I also need oxygen to function. Ah, there's nothing like a little fresh air!

Zoom in

Vessels that supply the heart

Coronary circulation involves the arterial system of blood vessels that supply oxygenated blood to the heart and the venous system that removes oxygen-depleted blood from it.

As a coronary artery, my job is to supply the heart with the oxygen it needs to keep beating.

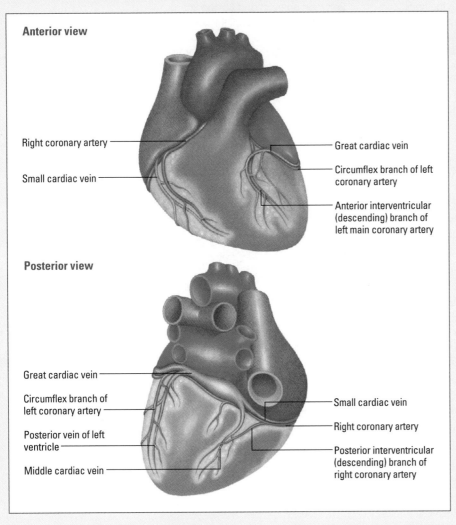

Anterior view

Right coronary artery

Small cardiac vein

Great cardiac vein

Circumflex branch of left coronary artery

Anterior interventricular (descending) branch of left main coronary artery

Posterior view

Great cardiac vein

Circumflex branch of left coronary artery

Posterior vein of left ventricle

Middle cardiac vein

Small cardiac vein

Right coronary artery

Posterior interventricular (descending) branch of right coronary artery

As time goes by . . .

Cardiovascular changes with aging

As a normal part of aging, the heart usually becomes slightly smaller. Contractile strength also declines, making the heart less efficient. In most people, resting cardiac output diminishes 30% to 35% by age 70.

Veins dilate and stretch with age, and coronary artery blood flow drops 35% between ages 20 and 60. The aorta becomes more rigid, causing systolic blood pressure to rise disproportionately higher than the diastolic, resulting in a widened pulse pressure.

Between ages 30 and 80, the left ventricular wall grows 25% thicker from its increased efforts to pump blood. Heart valves also become thicker from fibrotic and sclerotic changes. This can prevent the valves from closing completely, causing systolic murmurs.

Superficially speaking

The *cardiac veins* lie superficial to the arteries. The largest vein, the *coronary sinus*, opens into the right atrium. Most of the major cardiac veins empty into the coronary sinus, except for the *anterior cardiac veins*, which empty into the right atrium. (See *Cardiovascular changes with aging*.)

Quick quiz

1.   During systole the ventricles contract. This causes:
     A.   all four heart valves to close.
     B.   the AV valves to close and the semilunar valves to open.
     C.   the AV valves to open and the semilunar valves to close.
     D.   all four heart valves to open.

Answer: B. During systole, the pressure is greater in the ventricles than in the atria, causing the AV valves (the tricuspid and mitral valves) to close. The pressure in the ventricles is also greater than the pressure in the aorta and pulmonary artery, forcing the semilunar valves (the pulmonary and aortic valves) to open.

2.   The normal pacemaker of the heart is:
     A.   the SA node.
     B.   the AV node.
     C.   the ventricles.
     D.   the Purkinje fibres.

Answer: A. The SA node is the normal pacemaker of the heart, generating impulses 60 to 100 times per minute. The AV node is the secondary pacemaker of the heart (generating 40 to 60 beats per minute). The ventricles are the last line of defence (generating 20 to 40 beats per minute).

3.   The pressure the ventricular muscle must generate to overcome the higher pressure in the aorta refers to:
   A.   contractility.
   B.   preload.
   C.   blood pressure.
   D.   afterload.

Answer: D. Afterload is the pressure the ventricular muscle must generate to overcome the higher pressure in the aorta to get the blood out of the heart.

4.   The vessels that carry oxygenated blood back to the heart and left atrium are:
   A.   capillaries.
   B.   pulmonary veins.
   C.   pulmonary arteries.
   D.   superior and inferior venae cavae.

Answer: B. Oxygenated blood returns by way of venules and veins to the pulmonary veins, which carry it back to the heart's left atrium.

5.   The layer of the heart responsible for contraction is the:
   A.   myocardium.
   B.   pericardium.
   C.   endocardium.
   D.   epicardium.

Answer: A. The myocardium has striated muscle fibres that cause the heart to contract.

Scoring

☆☆☆   If you answered all five questions correctly, marvellous! You've got to the heart of the cardiovascular system.

☆☆   If you answered four questions correctly, great! We won't call you 'vein' if you're a little proud of yourself.

☆   If you answered fewer than four questions correctly, take heart! Re-circulate the information given in this chapter before moving on to the next.

Just the facts

In this chapter, you'll learn:
- the way in which blood cells develop
- functions of the different blood components
- the way in which blood cells clot
- blood groups and their significance.

A look at the haematological system

The haematological system consists of the blood and bone marrow. Blood delivers oxygen and nutrients to all tissues, removes wastes, and transports gases, blood cells, immune cells, antibodies and hormones throughout the body.

Living up to their potential

The haematological system manufactures new blood cells through a process called *haematopoiesis. Multipotential stem cells* in bone marrow give rise to five distinct cell types, called *unipotential stem cells*. Unipotential cells differentiate into one of the following four types of blood cells:
- eryhrocytes (the most common type)
- granulocytes
- agranulocytes
- platelets.
  (See *Tracing blood cell formation*, page 122.)

I've got multipotential!

Blood components

Blood consists of various formed elements, or *blood cells*, suspended in a fluid called *plasma*.

The RBCs of blood—and the WBCs and platelets, too

Formed elements in the blood include:
- red blood cells (RBCs), or erythrocytes
- white blood cells (WBCs), or leucocytes
- platelets, or thrombocytes.

RBCs and platelets function entirely within blood vessels; some WBCs remain in the blood while others can enter tissues.

Red blood cells

RBCs mainly transport oxygen to from the lungs to the body tissues. They contain *haemoglobin*, the oxygen-carrying substance that gives blood its red colour. When the red cells have given up the oxygen, they are capable of transporting some carbon dioxide back to the lungs for removal. However, carbon dioxide is mainly transported in the blood as bicarbonate.

The body manufactures billions of new RBCs like us every day. Almost 2 million a second!

The life and times of the RBC

RBCs have an average life span of 120 days. Bone marrow releases RBCs into circulation in immature form as *reticulocytes*. The reticulocytes mature into RBCs in about 1 day. The spleen removes old, worn-out RBCs from circulation.

A balance between removal and renewal

The rate of reticulocyte release usually equals the rate of old RBC removal. When RBC depletion occurs (e.g. with haemorrhage), the bone marrow increases reticulocyte production to maintain the normal RBC count. (See *Haematological changes with aging*, page 124.)

White blood cells

Five types of WBCs participate in the body's defence and immune systems. These five types of cells are classified as *granulocytes* (neutrophils, eosinophils and basophils) and *agranulocytes* (monocytes and lymphocytes).

Granulocytes

Granulocytes are a group of WBCs that contain granules in their cytoplasm. They can be subclassified into *neutrophils*, *eosinophils* and *basophils*. Neutrophils

Now I get it!

Tracing blood cell formation

Blood cells form and develop in the bone marrow by a process called *haematopoiesis*. This chart breaks down the process from when the five unipotential stem cells initiate from the multipotential stem cell until they mature into fully formed cells—either erythrocytes, granulocytes, agranulocytes or platelets.

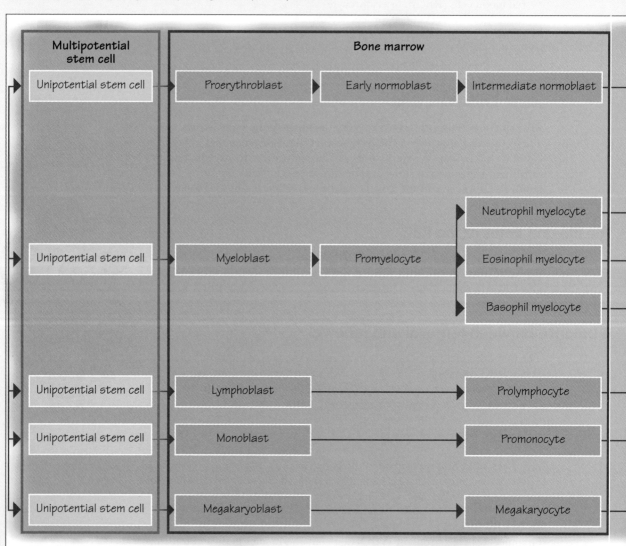

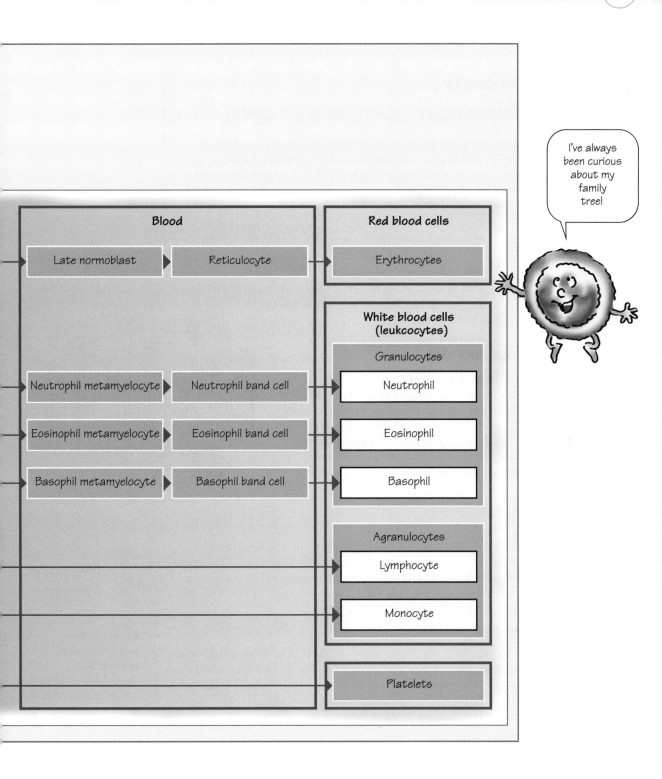

As time goes by . . .

Haematological changes with aging

As a person ages, fatty bone marrow replaces some of the body's active blood-forming marrow—first in the long bones and later in the flat bones. The altered bone marrow can't increase erythrocyte production as readily in response to such stimuli as hormones, anoxia, haemorrhage and haemolysis. Vitamin B_{12} absorption may also diminish with age, resulting in reduced erythrocyte mass and decreased haemoglobin levels and haematocrit (packed cell volume).

contain a multilobed nucleus, while nucleus of eosinophils and basophils are bilobed. Each cell type exhibits different properties and each is activated by different stimuli.

Swallowing up your enemies

Neutrophils, the most numerous granulocytes, account for 50% to 75% of circulating WBCs. These phagocytic cells engulf, ingest and digest foreign materials. They leave the bloodstream by passing through the capillary walls into the tissues (a process called *diapedesis*) and then migrate to and accumulate at infection sites. Neutrophils are the first cells to arrive at the site of injury.

Neutrophils are the first cells to arrive at the site of injury.

Making the band

Non-viable neutrophils form the main component of pus. Bone marrow produces their replacements, immature neutrophils called *bands*. In response to infection, bone marrow must produce many immature cells and release them into circulation, elevating the band count.

Allies against allergies

Eosinophils account for 0.3% to 7% of circulating WBCs. These granulocytes also migrate from the bloodstream by diapedesis but do so as a response to an allergic reaction. Eosinophils are involved in many inflammatory processes, especially those that arise from parasite infections and allergies. They release many chemicals involved in activating the immune response.

Memory jogger

To help yourself remember what **neutrophils** do, think of two other 'n' words: **n**umerous and **n**eutralise.

Fighting the flames

Basophils usually constitute fewer than 2% of circulating WBCs. They possess little or no phagocytic ability. Their cytoplasmic granules secrete *histamine* in response to certain inflammatory and immune stimuli. Histamine dilates the blood vessels making them more permeable, which eases the passage of fluids from the capillaries into body tissues.

Agranulocytes

WBCs in the agranulocyte category—*monocytes* and *lymphocytes*—lack visible cytoplasmic granules and have nuclei without lobes. (See *Comparing granulocytes and agranulocytes*, page 126.)

The few and the large

Monocytes, the largest of the WBCs, constitute only 1% to 9% of WBCs in circulation. Like neutrophils, monocytes are phagocytic and enter the tissues by diapedesis. Outside the bloodstream, monocytes enlarge and mature, becoming tissue *macrophages*.

We macrophages may be immobile at the moment, but at the first sign of inflammation, we're outta here!

Protection against infection

As macrophages, monocytes may roam freely through the body when stimulated by inflammation. Usually, they remain localized to organs and tissues. Collectively, they serve as components of the *reticuloendothelial system*, which defends the body against infection and disposes of cell breakdown products.

Fluid finders

Macrophages concentrate in structures that filter large amounts of body fluid, such as the liver, spleen and lymph nodes, where they defend against invading organisms. Macrophages are efficient *phagocytes*, cells that ingest microorganisms, cellular debris (including worn-out neutrophils) and necrotic tissue. When mobilised at an infection site, they become involved in the phagocytosis of cellular remnants and promote wound healing.

Last and, in fact, least (in size)

Lymphocytes, the second most common WBCs (20% to 43%), derive from stem cells in the bone marrow. There are three main types of lymphocytes:
- *T lymphocytes* directly attack an infected cell.
- *B lymphocytes* produce antibodies against specific antigens.
- *Natural killer cells* provide immune surveillance and primarily target virus-infected, and cancerous cells.

Now I get it!

Comparing granulocytes and agranulocytes

White blood cells (WBCs) are like soldiers fighting off the enemy. Each type of WBC fights a different enemy.

On the front line

Granulocytes, which comprise the 'platoons' of basophils, neutrophils and eosinophils, are the first forces 'marshalled' against invading foreign organisms.

Basophils release histamine in response to inflammatory and immune stimuli.

Neutrophils ingest foreign bodies.

Eosinophils ingest antigens and antibodies.

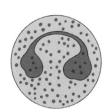

In the trenches

Agranulocytes, with 'platoons' of lymphocytes and monocytes, may roam freely on 'patrol' when inflammation is reported, but they mainly 'dig in' at structures that filter large amounts of fluid (such as the liver) and defend against invaders.

Lymphocytes ingest foreign material or produce antibodies.

Monocytes remove bacteria, cellular debris and necrotic tissue.

Platelets

Platelets are small, colourless, disc-shaped cytoplasmic fragments split from cells in bone marrow called *megakaryocytes*.

These fragments, which have a life span of approximately 10 days, perform three vital functions:
• initiating contraction of damaged blood vessels to minimise blood loss
• forming *haemostatic plugs* in injured blood vessels
• with plasma, providing materials that accelerate blood coagulation.

Blood clotting

Haemostasis is the complex process by which *platelets*, *plasma* and *coagulation factors* interact to control bleeding.

When cells like me are damaged, we release tissue factor, which activates the extrinsic portion of the coagulation system.

Stop the bleeding!

When a blood vessel ruptures, local *vasoconstriction* (decrease in the lumen diameter of blood vessels) and *platelet clumping* (aggregation) at the site of the injury initially help prevent haemorrhage. The damaged cells then release *tissue factor*, which activates the extrinsic pathway of the coagulation system.

A more long-term solution

However, formation of a more stable clot requires initiation of the complex clotting mechanisms known as the *intrinsic pathway*. This clotting system is activated by a protein, called Factor XII, one of 12 substances necessary for coagulation and derived from plasma and tissue.

Come together

The final result of coagulation is a *fibrin clot*, an accumulation of a fibrous, insoluble protein at the site of the injury. (See *How blood clots*, page 128.)

Coagulation factors

The materials that platelets and plasma provide work with *coagulation factors* to serve as *precursor compounds* in the clotting (coagulation) of blood.

12 Factors clotting

Designated by name and Roman numeral, these *coagulation factors* are activated in a chain reaction, each one in turn activating the next factor in the chain:
• Factor I, *fibrinogen*, is a high-molecular-weight protein synthesised in the liver and converted to fibrin during the coagulation cascade.
• Factor II, *prothrombin*, is a protein synthesised in the liver in the presence of vitamin K and converted to thrombin during coagulation.
• Factor III, *tissue factor*, is released from damaged tissue; it's required to initiate the second phase, the extrinsic pathway.
• Factor IV, consisting of *calcium ions*, is required throughout the entire clotting sequence.
• Factor V, or *proaccelerin*, is a protein that's synthesised in the liver and functions during the common pathway phase of the coagulation system.
• Factor VII, *proconvertin*, is a protein synthesised in the liver in the presence of vitamin K; it's activated by Factor III in the extrinsic system.
• Factor VIII, *antihaemophilic factor*, is a protein synthesised in the liver and required during the intrinsic phase of the coagulation system.

Now I get it!

How blood clots

When a blood vessel is severed or injured, three interrelated processes take place.

Constriction and aggregation

Immediately, the vessels affected by the injury contract (*constriction*), reducing blood flow. Also, platelets, stimulated by the exposed collagen of the damaged cells, begin to clump together (*aggregation*). Aggregation provides a temporary seal and a site for clotting to take place. The platelets release a number of substances that enhance constriction and aggregation.

Clotting pathways

Clotting, or *coagulation*—the transformation of blood from a liquid to a solid—may be initiated through two different pathways: the intrinsic pathway or the extrinsic pathway. The *intrinsic pathway* is activated when plasma comes in contact with damaged vessel surfaces. The *extrinsic pathway* is activated when tissue factor (a substance released by damaged endothelial cells) comes into contact with one of the clotting factors.

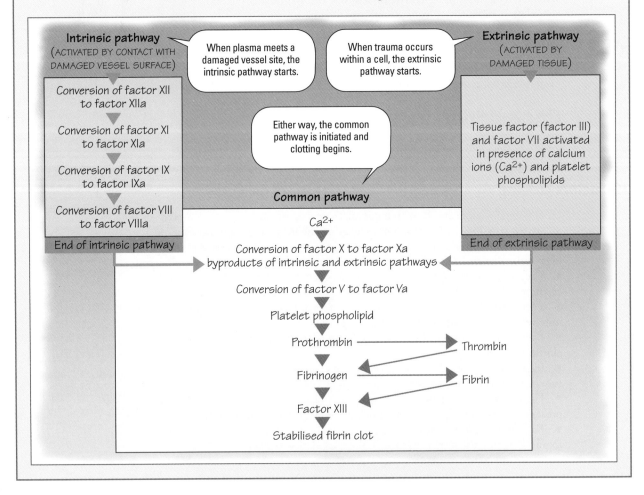

- Factor IX, *plasma thromboplastin component*, a protein synthesised in the liver in the presence of vitamin K, is required in the intrinsic phase of the coagulation system.
- Factor X, *Stuart-Prower factor*, is a protein synthesised in the liver in the presence of vitamin K; it's required in the common pathway of the coagulation system.
- Factor XI, *plasma thromboplastin antecedent*, is a protein synthesised in the liver and required in the intrinsic pathway.
- Factor XII, *Hageman factor*, is a protein required in the intrinsic pathway.
- Factor XIII, *fibrin stabilising factor*, is a protein required to stabilise the fibrin strands in the common pathway phase of the coagulation system.

Blood groups

Blood groups are determined by the presence or absence of genetically determined *antigens* or *agglutinogens* (glycoproteins) on the surface of RBCs. A, B and Rh are the most clinically significant blood antigens.

ABO groups

Testing for the presence of A and B antigens on RBCs is the most important system for classifying blood:
- Type A blood has A antigen on its surface.
- Type B blood has B antigen.
- Type AB blood has both A and B antigens.
- Type O blood has neither A nor B antigen.

Opposites don't attract

Plasma may contain *antibodies* that interact with these antigens, causing the cells to *agglutinate*, or combine into a mass. However, plasma can't contain antibodies to its red cell antigens or it would destroy these cells. Thus, type A blood has A antigen but no A antibodies; however, it does have B antibodies.

Making a match

Precise blood-typing and cross-matching (mixing and observing for agglutination of donor cells) are essential, especially for blood transfusions. A donor's blood must be compatible with a recipient's or the result can be fatal. The following blood groups are compatible:
- type A with type A or O
- type B with type B or O
- type AB with type A, B, AB or O
- type O with type O only.

Memory jogger

Blood types are easy to remember because they're named after the antigens they contain—**A** or **B** or both **A and B**—except for type **O**, which contains neither. The O serves as a nice visual reminder of that absence.

In reality, blood cells possess many different antigens in addition to antigen A and antigen B. Blood is always cross-matched so that the recipient receives the same type of blood as their own.

(*See Reviewing blood type compatibility.*)

Now I get it!

Reviewing blood type compatibility

Precise blood typing and cross-matching can prevent the transfusion of incompatible blood, which can be fatal. Usually, typing the recipient's blood and cross-matching it with available donor blood take less than 1 hour.

Making a match

Agglutinogen (an antigen in red blood cells) and *agglutinin* (an antibody in plasma) distinguish the four ABO blood groups. This chart shows ABO compatibility from the perspectives of the recipient and the donor.

Blood group	Antibodies present in plasma	Compatible RBCs	Compatible plasma
Recipient			
O	Anti-A and anti-B	O	O, A, B, AB
A	Anti-B	A, O	A, AB
B	Anti-A	B, O	B, AB
AB	Neither anti-A nor anti-B	AB, A, B, O	AB
Donor			
O	Anti-A and anti-B	O, A, B, AB	O
A	Anti-B	A, AB	A, O
B	Anti-A	B, AB	B, O
AB	Neither anti-A nor anti-B	AB	AB, A, B, O

Blood typing is an important step before a transfusion.

Rh typing

Another important type of blood grouping relates to the Rhesus (Rh) factor. Rh typing determines whether Rh factor is present or absent in blood. This enables blood to be typed as Rh positive or Rh negative.

Positive and negative types

Typically, blood contains the Rh antigen. Blood with the Rh antigen is Rh-positive; blood without the Rh antigen is Rh-negative. Anti-Rh antibodies can appear only in a person who has become sensitised. Anti-Rh antibodies can appear in the blood of an Rh-negative person after entry of Rh-positive RBCs in the bloodstream—for example, from transfusion of Rh-positive blood. An Rh-negative female who carries an Rh-positive fetus may also acquire anti-Rh antibodies.

Quick quiz

1. The component of blood that triggers defence and immune responses is the:
    A. WBC.
    B. platelet.
    C. RBC.
    D. haemoglobin.

Answer: A. Because of their phagocytic capabilities, WBCs serve as the body's first line of cellular defence against foreign organisms.

2. The complex process by which platelets, plasma and coagulation factors interact to control bleeding is called:
    A. phagocytosis.
    B. haematopoiesis.
    C. haemostasis.
    D. diapedesis.

Answer: C. Haemostasis is achieved through a three-part process: vasoconstriction, platelet aggregation and coagulation.

3. Blood cells form and develop in the:
    A. kidneys.
    B. liver.
    C. pancreas.
    D. bone marrow.

Answer: D. Multipotential stem cells in the bone marrow give rise to five distinct cell types called unipotential stem cells. Each of these stem cells can differentiate into an erythrocyte, a granulocyte, an agranulocyte or a platelet.

4.   The most numerous type of granulocytes are the:
A.   bands.
B.   neutrophils.
C.   eosinophils.
D.   basophils.

Answer: B. Neutrophils are the most numerous granulocytes, accounting for 50% to 75% of circulating WBCs.

5.   Blood groups are determined by testing for A and B antigens on the:
A.   red blood cells.
B.   white blood cells.
C.   platelets.
D.   thrombocytes.

Answer: A. Blood groups are determined by the presence or absence of antigens or agglutinogens on the surface of RBCs.

Scoring

★★★ If you answered all five questions correctly, wonderful! You're clearly thinking haematologically!

★★ If you answered four questions correctly, great. You've coagulated all the information in this chapter into a solid understanding of blood.

★ If you answered fewer than four questions correctly, be more positive (or A positive or AB positive). Just read the chapter again and give the quiz another go.

Just the facts

In this chapter, you'll learn:

♦ organs and tissues that make up the immune system

♦ functions of the immune system

♦ the body's response when the immune system fails.

A look at the immune system

The immune system defends the body against invasion by harmful organisms and toxins.

Lymphoid rules

Organs and tissues of the immune system are referred to as 'lymphoid' because they're all involved with the growth, development and dissemination of lymphocytes, one type of white blood cell (WBC). (See *Organs and tissues of the immune system*, page 134.)

The immune system has three major components:

• central lymphoid organs and tissue
• peripheral lymphoid organs and tissue
• accessory lymphoid organs and tissue.

The immune system defends the body against invasion by harmful organisms and toxins.

Blood relatives

Although the immune system and blood are distinct entities, they're closely related. Their cells share a common origin in the bone marrow, and the immune system uses the bloodstream to transport its active components to where they are required.

> The immune system defends the body against invasion by harmful organisms and toxins.

Body shop

Organs and tissues of the immune system

The immune system includes organs and tissues in which lymphocytes predominate as well as cells that circulate in the blood. This illustration shows central, peripheral and accessory lymphoid organs and tissue.

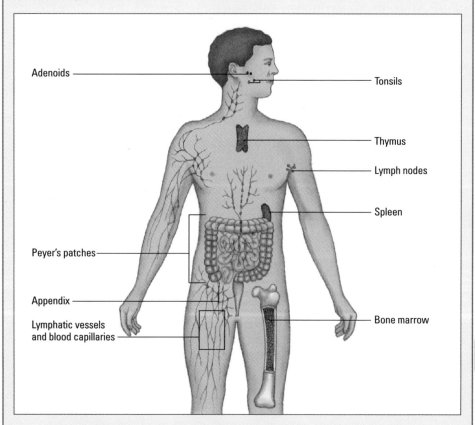

Adenoids

Tonsils

Thymus

Lymph nodes

Spleen

Peyer's patches

Appendix

Lymphatic vessels and blood capillaries

Bone marrow

We're T and B cells, the two major types of lymphocytes.

Central lymphoid organs and tissues

The bone marrow and the thymus each play a role in the development of B cells and T cells—the two major types of *lymphocytes*.

Bone marrow

The *bone marrow* contains stem cells, which can develop into any of several different cell types. Such cells are *multipotential*, meaning they're capable of taking many forms. The cells of the immune system and the blood develop from stem cells in a process called *haematopoiesis*.

To be B or to be T? That is the question . . .

Soon after their differentiation from other stem cells, some of the cells destined to become immune system cells serve as sources for *lymphocytes*; other cells of this differentiated group develop into *phagocytes* (cells that ingest microorganisms). Those that become lymphocytes are further differentiated to become either *B cells* (which mature in the bone marrow) or *T cells* (which travel to the thymus and mature there).

T is better to receive

B cells and T cells are distributed throughout the lymphoid organs, especially the lymph nodes and spleen. T and B lymphocytes have special receptors that respond to specific antigen molecule shapes. B cells also produce special proteins known as *antibodies* which bind specific *antigens* (foreign proteins against which the antibodies are generated) to help neutralise their effects. Antibodies attack pathogens or direct other cells, such as phagocytes, to attack for them.

Thymus

In the foetus and infant, the *thymus* is a two-lobed mass of lymphoid tissue that's located in the upper anterior part of the chest cavity in the mediastinum. The thymus helps form T lymphocytes (also known as *T cells*) for several months after birth. After this time, it has no function in the body's immunity. It reaches maximum size at puberty and then begins to atrophy until only a remnant remains in adults.

Basic training

In the thymus, T cells undergo a process called *T-cell education*, in which the cells are 'trained' to recognise other cells from the same body (self cells) and distinguish them from all other cells (nonself cells). There are several types of T cells, each with a specific function:

- memory T cells
- helper T cells
- regulatory T cells
- suppressor T cells
- natural killer (cytotoxic T cells).

Peripheral lymphoid organs and tissues

Peripheral structures of the immune system include the lymph nodes, the lymphatic vessels and the spleen.

I'm a multipotential cell. That means I'm capable of wearing many different hats.

Memory jogger

To recall where lymphocytes mature, think, 'B in B, and T in T'. **B** cells mature in the **B**one marrow, and **T** cells mature in the **T**hymus.

Lymph nodes

The *lymph nodes* are small, oval-shaped structures located along a network of *lymph channels*. Most abundant in the head, neck, axillae, abdomen, pelvis, and groin, lymph nodes help remove and destroy *antigens* (substances capable of triggering an immune response) that circulate in the blood and lymph.

Fully furnished compartments

Each lymph node is enclosed in a fibrous capsule. From this capsule, bands of connective tissue extend into the node and divide it into three compartments:

 The *superficial cortex* contains follicles made up predominantly of B cells.

 The *deep cortex* and interfollicular areas consist mostly of T cells.

The *medulla* contains numerous plasma cells derived from B cells that actively secrete *immunoglobulins*.

Lymphatic vessels

Lymph is a clear fluid that bathes the body tissues. It contains a liquid portion, which resembles blood plasma, as well as WBCs (mostly lymphocytes and macrophages) and may also contain antigens. Collected from body tissues, lymph seeps into *lymphatic vessels* across the vessels' thin walls. (See *Lymphatic vessels and lymph nodes*.)

Carried into the cavities . . .

Afferent lymphatic vessels carry lymph into the *subcapsular sinus* (or cavity) of the lymph node. From here, lymph flows through cortical sinuses and smaller radial medullary sinuses. Phagocytic cells in the deep cortex and medullary sinuses attack any foreign antigens carried in lymph. The antigens also may be trapped in the follicles of the superficial cortex. These processes essentially clean the lymph.

. . . and coming out cleansed

Cleansed lymph leaves the node through *efferent lymphatic vessels* at the *hilum* (a depression at the exit or entrance of the node). These vessels drain into *lymph node chains* that, in turn, empty into large lymph vessels, or trunks, which drain into the subclavian vein of the vascular system.

Getting security clearance

Usually, lymph travels through more than one lymph node because numerous nodes line the lymphatic channels that drain a particular region. For example, axillary nodes (located under the arm) filter drainage from the arms, and femoral nodes (in the inguinal region) filter drainage from the legs. This arrangement limits the movement of organisms that enter peripheral areas from migrating unchallenged to central areas.

Memory jogger

Afferent means to **bring to** and efferent means to **lead away**. Therefore, it's easy to remember that the afferent lymphatic vessels **bring** lymph into the sinuses and the efferent lymphatic vessels **lead it away**.

Zoom in

Lymphatic vessels and lymph nodes

Lymphatic tissues are connected by a network of thin-walled drainage channels called *lymphatic vessels*. Resembling veins, the afferent lymphatic vessels carry lymph into lymph nodes; lymph slowly filters through the node and is collected into efferent lymphatic vessels.

It can check in but it can't check out

Lymphatic capillaries are located throughout most of the body. Wider than blood capillaries, they permit interstitial fluid to flow into them but not out.

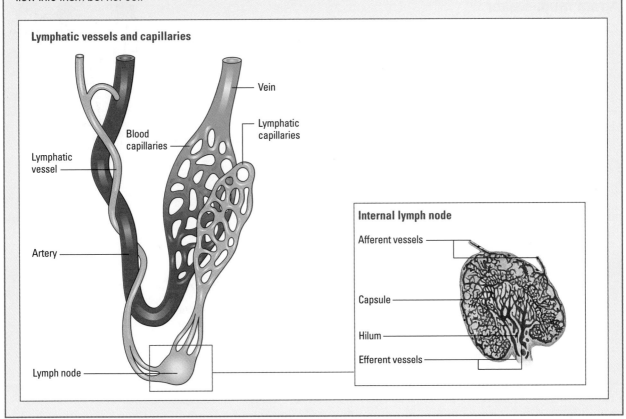

Lymphatic vessels and capillaries

Vein

Lymphatic capillaries

Blood capillaries

Lymphatic vessel

Artery

Lymph node

Internal lymph node

Afferent vessels

Capsule

Hilum

Efferent vessels

Spleen

Located in the left upper quadrant of the abdomen beneath the diaphragm, the *spleen* is a dark red, oval structure that's approximately the size of a fist and is the largest lymphatic organ. Bands of connective tissue from the dense fibrous capsule surrounding the spleen extend into the spleen's interior.

The white and the red

The interior, called the *splenic pulp*, contains white and red pulp. *White pulp* contains compact masses of lymphocytes surrounding branches of the splenic artery. *Red pulp* consists of a network of blood-filled *sinusoids*, supported by a framework of reticular fibres and mononuclear phagocytes, along with some lymphocytes and plasma cells.

A real multi-tasker

The spleen has several functions:
• Its phagocytes engulf and break down worn-out red blood cells (RBCs), causing the release of haemoglobin, which then breaks down into its components. These phagocytes also selectively retain and destroy damaged or abnormal RBCs and cells with large amounts of abnormal haemoglobin.
• The spleen filters and removes bacteria and other foreign substances that enter the bloodstream; these substances are promptly removed by splenic phagocytes.
• Splenic phagocytes interact with lymphocytes to initiate an immune response.
• The spleen stores blood and 20% to 30% of platelets.
• In the foetus, the spleen produces RBCs.
• If the spleen is removed due to disease or trauma, the liver and bone marrow assume its function.

Accessory lymphoid organs and tissues

The *tonsils, pharyngeal tonsils, appendix* and *Peyer's patches* remove foreign debris in much the same way lymph nodes do. They're located in areas in which microbial access is more likely, such as the nasopharynx (tonsils and pharyngeal tonsils) and the abdomen (appendix and Peyer's patches).

Immune system function

Immunity refers to the body's capacity to resist invading organisms and toxins, thereby preventing tissue and organ damage. The immune system is designed to recognise, respond to, and eliminate antigens, including bacteria, fungi,

> The spleen is a dark red, oval structure that's approximately the size of a fist and is the largest lymphatic organ.

viruses and parasites. It also preserves the body's internal environment by scavenging dead or damaged cells and patrolling for antigens.

Strategic moves

To perform these functions efficiently, the immune system uses three basic strategies:
- protective surface phenomena
- general host defences
- specific immune responses.

Protective surface phenomena

Strategically placed physical, chemical and mechanical barriers work to prevent the entry of potentially harmful organisms.

The forward guard

Intact *skin* and *mucous membranes* provide the first line of defence against microbial invasion, preventing attachment of microorganisms. Skin *desquamation* (normal cell turnover) and low pH (due to acidic secretions) further impede bacterial colonisation. Seromucous surfaces are protected by antibacterial substances—for instance, the enzyme *lysozyme*, which is found in tears, saliva, sweat and nasal secretions.

Breathe easy . . .

In the respiratory system (the easiest part of the body for microorganisms to enter), *nasal hairs* and *turbulent airflow* through the nostrils filter out foreign materials. Nasal secretions contain an immunoglobulin that discourages microbe adherence. Also, a mucous layer, which is continuously sloughed off and replaced, lines the respiratory tract and provides additional protection. Cilia, on the cells of the upper respiratory tract, direct mucous and bacteria to the back of the throat to be swallowed.

. . . and swallow

In the GI tract, bacteria are mechanically removed by saliva, swallowing, peristalsis and defecation. In addition, the low pH of gastric secretions is *bactericidal* (bacteria-killing), rendering the stomach virtually free from live bacteria.

The remainder of the GI system is protected through *colonisation resistance*, in which resident bacteria prevent other microorganisms from permanently making a home.

No colonisation allowed

The urinary system is sterile except for the distal end of the urethra and the urinary meatus. Urine flow, low urine pH, immunoglobulin and, in men,

I'm a resident bacterium. I live in harmony with the body without causing disease, but I keep other microorganisms from colonising my patch.

the bactericidal effects of *prostatic fluid* work together to impede bacterial colonisation. A series of sphincters also inhibits bacterial migration.

General host defences

When an antigen penetrates the skin or mucous membrane, the immune system launches nonspecific cellular responses in an effort to identify and remove the invader.

Raising the red flag

The first of the nonspecific responses against an antigen, the *inflammatory response*, involves vascular and cellular changes, including the production and release of such chemical substances as heparin, histamine and kinin. These changes eliminate dead tissue, microorganisms, toxins and inert foreign matter. (See *Understanding the inflammatory response*.)

Inflammatory response rousers

Granular leucocytes play a big role in the inflammatory response:
- *Neutrophils*, which are produced in the bone marrow, are the most numerous polymorphonuclear leucocytes. They increase dramatically in number in response to infection and inflammation. They're the main constituent of pus and are highly mobile. Neutrophils are attracted to areas of inflammation. They engulf, digest and dispose of invading organisms through a process called *phagocytosis*.
- *Eosinophils*, found in large numbers in the respiratory system and GI tract, multiply in allergic and parasitic disorders. Although their phagocytic function isn't clearly understood, evidence suggests that they participate in host defence against parasites.
- *Basophils* and *mast cells* also function in immune disorders. Basophils circulate in peripheral blood, whereas mast cells remain in the tissues— predominantly those of the lungs, intestines and skin. (Mast cells are not blood cells.) Both cells have surface receptors for immunoglobulin (Ig) E. When their receptors are cross-linked by an IgE antigen complex, they release mediators characteristic of the allergic response.

Specific immune responses

All foreign substances elicit the same general host defences. In addition, particular microorganisms or molecules activate specific immune responses and can involve specialised sets of immune cells. Specific responses, classified as either *humoral immunity* or *cell-mediated immunity*, are produced by lymphocytes (B cells and T cells).

Now I get it!

Understanding the inflammatory response

The inflammatory response helps the body return to homeostasis after a wound occurs. Its primary function is to bring phagocytic cells (neutrophils and monocytes) to the inflamed area to destroy bacteria and rid the tissue spaces of dead and dying cells so that tissue repair can begin.

Inflammation produces four cardinal signs: redness, swelling, heat and pain. The first three signs result from local vasodilation, fluid leakage into the extravascular space and blockage of lymphatic drainage. The fourth results from tissue space distention caused by swelling and pressure and from chemical irritation of nociceptors (pain receptors).

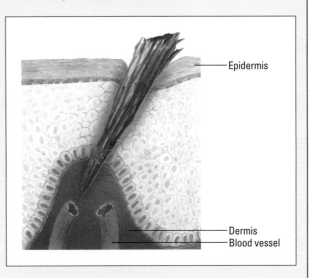

Epidermis

Dermis
Blood vessel

Humoral immunity

In this response, an invading antigen causes B cells to divide and differentiate into plasma cells. Each plasma cell, in turn, produces and secretes large amounts of antigen-specific immunoglobulins into the bloodstream.

The Ig guard

Each of the five types of *immunoglobulins* (IgA, IgD, IgE, IgG and IgM) serves a particular function:
- IgA, IgG and IgM guard against viral and bacterial invasion.
- IgD acts as an antigen receptor of B cells.
- IgE causes an allergic response.

'Y' we do it?

Immunoglobulins have a special molecular structure that creates a Y shape. The upper fork of the Y is designed to attach to a particular antigen; the lower stem enables the immunoglobulin to link with other structures in the immune system. Depending on the antigen, immunoglobulins can work in one of several ways:

My job is to produce antibodies. They attack the antigens for me.

- They can disable certain bacteria by linking with toxins that the bacteria produce. This disables the toxin and prevents it from binding to tissues and causing disease.
- They can *opsonise* (coat) bacteria, making them targets for scavenging by phagocytosis. (See *How macrophages accomplish phagocytosis.*)
- Most commonly, they can link to antigens, causing the liver to produce and circulate proteins called *complement*.

A break in the action

After the body's initial exposure to an antigen, a time lag occurs during which little or no antibody can be detected. During this time, the B cell recognises the antigen and the sequence of division, differentiation and antibody formation begins.

First response

The *primary antibody response* occurs 4 to 10 days after first-time antigen exposure. This primarily involves the production of IgM and IgG antibodies. During this response, IgG levels gradually increase and then quickly dissipate.

Second response: hit 'em hard, hit 'em fast

Subsequent exposure to the same antigen initiates a *secondary antibody response*. In this response, memory B cells manufacture antibodies (now mainly IgG), achieving peak levels in 1 to 2 days. These elevated levels persist for months and then fall slowly. Thus, the secondary antibody response is faster, more intense and more persistent than the primary response. This response intensifies with each subsequent exposure to the same antigen.

Getting complex

After the antibody reacts to the antigen, an *antigen-antibody complex* forms. The complex serves several functions. First, a macrophage processes the antigen and presents it to antigen-specific B cells. Then the antibody activates the complement system, which assists in destroying the antigen.

Complement system

The *complement system* is activated by a tissue injury or antigen-antibody reactions. It bridges humoral and cell-mediated immunity and attracts phagocytic neutrophils and macrophages to the antigen site.

Working together

Indispensable to the humoral immune response, the complement system consists of about 25 diverse proteins that 'complement' the work of antibodies by aiding phagocytosis or destroying bacterial cells (through puncture of their cell membranes).

I like the direct approach. Sometimes I attack antigens myself.

You thought I was slow in the first round? Well, just wait. I always come back faster and meaner in the second.

Now I get it!

How macrophages accomplish phagocytosis

Microorganisms and other antigens that invade the skin and mucous membranes are removed by *phagocytosis*, a defence mechanism carried out by macrophages (mononuclear leucocytes) and neutrophils (polymorphonuclear leucocytes). Here's how macrophages accomplish phagocytosis.

Chemotaxis
Chemotactic factors attract macrophages to the antigen site.

Microorganism

Chemotactic factors

Macrophage

Opsonisation
The antibody (immunoglobulin G) or complement fragment coats the microorganism, enhancing macrophage binding to the antigen, now called an *opsinogen.*

Opsonised microorganism

Ingestion
The macrophage extends its membrane around the opsonised microorganism, engulfing it within a vacuole (*phagosome*).

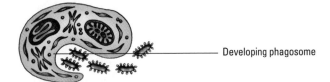

Developing phagosome

Digestion
As the phagosome shifts away from the cell periphery, it merges with lysosomes, forming a *phagolysosome*, where antigen destruction occurs.

Phagolysosome

Release
When digestion is complete, the macrophage expels digestive debris, including lysosomes, prostaglandins, complement components and interferon, which continue to mediate the immune response.

Digestive debris

A cascade effect

Complement proteins travel in the bloodstream in an inactive form. When the first complement substance is triggered (typically by an antibody interlocked with an antigen), it sets in motion a ripple effect. As each component is activated in turn, it acts on the next component in a sequence of carefully controlled steps called the *complement cascade*.

Attack mode

This cascade leads to the creation of the *membrane attack complex*. Inserted into the membrane of the target cell, this complex creates a channel through which fluid contents of the cells can escape causing lysis.

Other benefits flow from the complement cascade

By-products of the complement cascade also enhance:
• the inflammatory response (resulting from release of chemicals or cytokines from mast cells and basophils)
• stimulation and attraction of neutrophils (which participate in phagocytosis)
• coating of target cells by C3b (an activate fragment of the complement protein C3), making them attractive to phagocytes.

Cell-mediated immunity

Cell-mediated immunity protects the body against viral and fungal infections by inactivating the antigen and provides resistance against transplanted cells and tumour cells.

Ever vigilant

In this immune response, a macrophage processes the antigen, which is then presented to T cells. Some T cells become sensitised and inactivate the antigen; others release *lymphokines*, which activate macrophages that inactivate the antigen. Sensitised T cells then travel through the blood and lymphatic systems, providing ongoing surveillance in their quest for specific antigens. Other T cells produce cytotoxins which destroy both infected and cancer cells. (See *Immune response to bacterial invasion*.)

The great communicators

Cytokines are low-molecular-weight proteins involved in the communication between macrophages and lymphocytes. These proteins are responsible for inducing and regulating many immune and inflammatory responses. Cytokines include colony-stimulating factors, interferons, interleukins, tumour necrosis factors and transforming growth factor. They're an important part of a well-functioning immune system.

The complement cascade plays a crucial role in the inflammatory response.

Now I get it!

Immune response to bacterial invasion

Invasion of a foreign substance can trigger two types of immune responses—antibody-mediated (humoral) and cell-mediated immunity:

- In *humoral* immunity, antigens stimulate B cells to differentiate into plasma cells and produce circulating antibodies that disable bacteria and viruses before they can enter host cells.
- In *cell-mediated* immunity, T cells move directly to attack invaders. Three T-cell subgroups trigger the response to infection. Helper T cells spur B cells to manufacture antibodies. Suppressor T cells regulate T and B types of immune response. Cytotoxic T cells attack cells that express abnormal or foreign antigens and produce lymphokines (proteins that induce the inflammatory response and mediate the delayed hypersensitivity reaction).

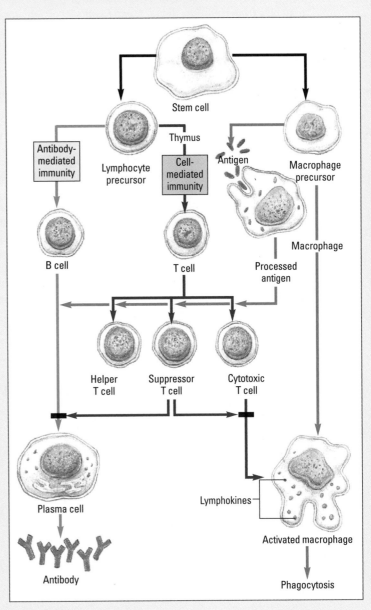

Immune system malfunction

Because of their complexity, the processes involved in host defence and immune response may malfunction. When the body's defences are exaggerated, misdirected, or either absent or depressed, the result may be a *hypersensitivity disorder*, *autoimmunity* or *immunodeficiency*, respectively.

Hypersensitivity disorders

An exaggerated or inappropriate immune response may lead to various hypersensitivity disorders.

Typing them out

Such disorders are classified as type I through type IV, depending on which immune system activity causes tissue damage, although some overlap exists:

- Type I disorders are *anaphylactic (acute hypersensitivityor allergic IgE mediated) reactions.* Examples of type I disorders include systemic anaphylaxis, hay fever (seasonal allergic rhinitis), reactions to insect stings, some food and drug reactions, some cases of urticaria and infantile eczema.
- Type II disorders are *cytotoxic (cytolytic, complement-dependent cytotoxicity) reactions.* Examples of type II disorders include Goodpasture's syndrome, autoimmune haemolytic anaemia, transfusion reactions, haemolytic disease of the neonate, myasthenia gravis and some drug reactions.
- Type III disorders are *immune complex disease reactions.* Examples of type III disorders are reactions associated with such infections as hepatitis B and bacterial endocarditis; cancers, in which a serum sickness-like syndrome may occur and autoimmune disorders such as systemic lupus erythematosus. This hypersensitivity reaction may also follow drug or serum therapy.
- Type IV disorders are *delayed (cell-mediated) hypersensitivity reactions.* Type IV disorders include tuberculin reactions, contact hypersensitivity (latex allergy) and sarcoidosis (nodular collections of inflammatory cells).

Autoimmune disorders

Autoimmune disorders are marked by an abnormal immune response to one's own tissue.

Hypersensitivity stems from an exaggerated or inappropriate immune response.

Diffusion can lead to confusion

Autoimmunity leads to a sequence of tissue reactions and damage that may produce diffuse systemic signs and symptoms. Among the autoimmune disorders are type 1 diabetes mellitus, rheumatoid arthritis, juvenile rheumatoid arthritis, psoriatic arthritis, ankylosing spondylitis, Sjögren syndrome, multiple sclerosis, autoimmune pancreatitis and lupus erythematosus. (See *Immunological changes with aging*.)

Immunodeficiency

Immunodeficiency disorders are caused by an absent or a depressed immune response in various forms.

Unfortunately, no deficiency of immunodeficiency disorders

Immunodeficiency disorders include X-linked infantile hypogammaglobulinaemia, common variable immunodeficiency, DiGeorge syndrome, acquired immunodeficiency syndrome, chronic granulomatous disease, ataxia-telangiectasia, severe combined immunodeficiency disease and complement deficiencies.

Immunodeficiency disorders occur when the immune system is depressed or on a downward slide.

As time goes by . . .

Immunological changes with aging

Immune function starts declining at sexual maturity and continues declining with age. During this decline, the immune system begins losing its ability to differentiate between self and nonself, and the incidence of autoimmune disease increases. The immune system also begins losing its ability to recognise and destroy mutant cells, which may account for the increase in cancer among older people.

External factors, such as nutritional status and exposure to chemical and environmental pollution and ultraviolet radiation, can also affect immune status.

Decreased antibody response in older people makes them more susceptible to infection. Tonsillar atrophy and lymphadenopathy commonly occur.

Total and differential leucocyte counts don't change significantly with age. However, some people over age 65 may exhibit a slight decrease in leucocyte count. When this happens, the number of B cells and total lymphocytes decreases, and T cells decrease in number and become less effective. Also, the sizes of the lymph nodes and spleen reduce slightly.

Quick quiz

1. Stem cells are multipotential and develop into other types of cells through the process of:
    A. chemotaxis.
    B. phagocytosis.
    C. haematopoiesis.
    D. opsonisation.

Answer: C. Haematopoiesis is the formation of blood cells, which occurs in the bone marrow.

2. Cleansed lymph leaves the lymph nodes through:
    A. afferent lymphatic vessels.
    B. efferent lymphatic vessels.
    C. lymphatic capillaries.
    D. blood capillaries.

Answer: B. Efferent lymphatic vessels drain into lymph node chains, then into large lymph vessels and, finally, into the subclavian vein.

3. During which phase of phagocytosis does a macrophage engulf an opsonised microorganism within a vacuole?
    A. Chemotaxis
    B. Opsonisation
    C. Ingestion
    D. Digestion

Answer: C. During ingestion, the macrophage extends its membrane around the microorganism, engulfing it within a vacuole and forming a phagosome.

Scoring

☆☆☆   If you answered all three questions correctly, impressive! You're definitely on top of our immunological challenges!

☆☆   If you answered two questions correctly, brilliant! You've proved your multipotential!

☆   If you answered only one question correctly, you might say you have an immunodeficiency. Take some vitamin C, and read this chapter again in the morning!

Just the facts

In this chapter, you'll learn:

♦ structures of the respiratory system and their functions

♦ the processes of inspiration and expiration

♦ the way in which gas exchange takes place

♦ problems with the nervous, musculoskeletal and pulmonary systems that can affect breathing

♦ the role of the lungs in acid-base balance.

A look at the respiratory system

The respiratory system maintains the exchange of oxygen and carbon dioxide in the lungs and tissues. It also helps regulate the body's acid-base balance. Functionally, the respiratory system is composed of a conducting region and a respiratory region. The conducting region consists of the continuous passageway that transports air in and out of the lungs (nose, pharynx, larynx, trachea, bronchi and terminal bronchioles). The respiratory region, composed of the respiratory bronchioles, alveolar ducts and alveoli, is the site of gas exchange.

The respiratory system can also be classified in terms of the upper and lower respiratory tracts. The lower respiratory tract is contained largely within the thoracic cavity. The thoracic cavity functions in the provision of protection of a number of components of the respiratory system and is involved in the mechanics of breathing.

Hey, did you know we have two regions?

Upper respiratory tract

The upper respiratory tract consists primarily of the nose (nostrils and nasal passages), mouth, nasopharynx, oropharynx, laryngopharynx and larynx. These structures filter, warm and humidify inspired air. The pharynx

connects the nasal cavity to the larynx and also the oesophagus.
(See *Structures of the respiratory system.*)

Nostrils and nasal passages

Air enters the body through the nostrils (*external nares*). In the external
nares, small hairs known as *vibrissae* filter out dust and large foreign
particles. Air then passes into the two nasal passages, which are separated
by the *septum*. Cartilage forms the anterior walls of the nasal passages;
bony structures (*conchae* or *turbinates*) form the posterior walls.

Just passing through

The *conchae* assists in the warming and humidification of air before
it passes into the nasopharynx. Their mucus layer also traps finer
foreign particles, which the *cilia* (small, hairlike projections) carry
to the pharynx to be swallowed.

Sinuses and nasopharynx

The four paranasal sinuses are located in the frontal, sphenoid and
maxillary bones. The sinuses provide speech resonance.

Air passes from the nasal cavity into the muscular nasopharynx through
a pair of posterior openings in the nasal cavity (*choanae*) that remain
constantly open. The nasopharynx is located behind the nose and above the
throat.

Oropharynx and laryngopharynx

The oropharynx is the posterior wall of the mouth. It connects the
nasopharynx and the laryngopharynx. The laryngopharynx extends to the
oesophagus and larynx.

Larynx

The larynx contains the vocal cords and connects the pharynx with the
trachea. Muscles and cartilage form the walls of the larynx, including the
large, shield-shaped thyroid cartilage located in the anterior superior position
in the neck. The anterior part of the thyroid cartilage forms the Adam's
apple.

Situated in front of the entrance to the larynx lies the epiglottis, a lid-like
flap of connective tissue that is attached to the root of the tongue. When
resting, the epiglottis allows air to pass through the larynx and into the rest of
the respiratory system. When swallowing, it covers the entrance to the larynx
preventing food and drink from entering the windpipe.

It's nothing to sneeze at. These involuntary defence mechanisms help protect the respiratory system from infection and foreign-body inhalation.

Body shop

Structures of the respiratory system

The principal structures of the respiratory system can be divided into the upper and lower respiratory tracts or into a conducting and respiratory portion.

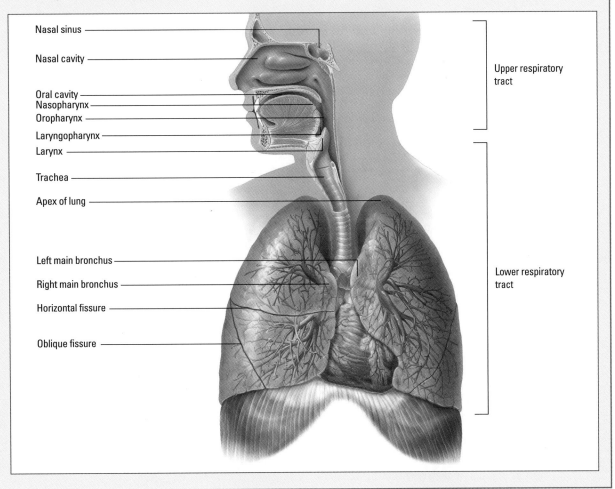

Nasal sinus

Nasal cavity

Oral cavity
Nasopharynx
Oropharynx

Laryngopharynx

Larynx

Trachea

Apex of lung

Left main bronchus

Right main bronchus

Horizontal fissure

Oblique fissure

Upper respiratory tract

Lower respiratory tract

Lower respiratory tract

The lower respiratory tract consists of the trachea, bronchi and lungs. The air passages are lined with mucous membrane composed mainly of ciliated epithelium. Cilia constantly clean the tract and carry foreign matter upwards for swallowing or expectoration.

Trachea

The trachea (windpipe) extends from the laryngopharynx at the level of the *cricoid cartilage* at the top to the carina (also called the *tracheal bifurcation*). C-shaped cartilage rings reinforce and protect the trachea to prevent it from collapsing. The carina is a ridge-shaped structure at the level of T6 or T7. The carina possesses sensory nerve endings which cause coughing if food or water is inhaled accidently.

Bronchi

The primary bronchi begin at the carina. The right primary bronchus—shorter, wider and more vertical than the left—supplies air to the right lung. The left primary bronchus delivers air to the left lung. Along with blood vessels, nerves, and lymphatics, the primary bronchi enter the lungs at the *hilum*. Located behind the heart, the hilum is a slit on the lung's medial surface.

> Bronchi branch out—from lobar bronchi to segmental bronchi to bronchioles.

Secondary bronchi

Each primary bronchus divides to form secondary bronchi. In each lung, one secondary bronchus goes into each lobe which means that the right lung has three secondary bronchi and the left lung has two.

Branching out

Each lobar bronchus enters a lobe in each lung. Within its lobe, each of the lobar bronchi branches into segmental bronchi (tertiary bronchi). The segments continue to branch into smaller and smaller bronchi, finally branching into bronchioles.

The larger bronchi consist of cartilage, smooth muscle and epithelium. As the bronchi become smaller, they lose cartilage and then smooth muscle. Ultimately, the smallest bronchioles consist of just a single layer of epithelial cells.

Respiratory bronchioles

Each bronchiole includes terminal bronchioles and the alveolar sac—the chief respiratory unit for gas exchange. (See *A close look at a pulmonary airway.*)

Within the acinus, terminal bronchioles branch into yet smaller respiratory bronchioles. The respiratory bronchioles feed directly into alveoli at sites along their walls.

Zoom in

A close look at a pulmonary airway

The respiratory unit (acinus) consists of the respiratory bronchiole, alveolar duct and sac and alveoli. Gas exchange occurs rapidly in the alveoli, in which oxygen from inhaled air diffuses into the blood and carbon dioxide diffuses from the blood into exhaled air.

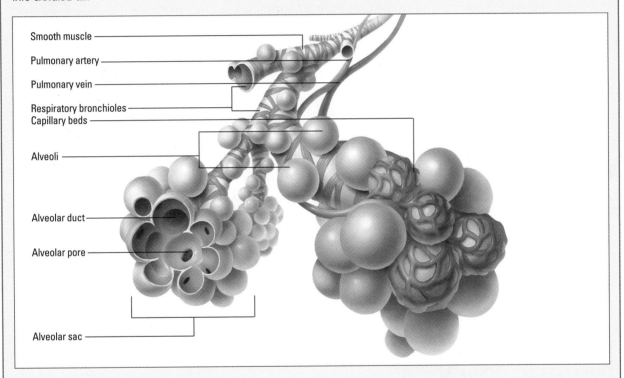

Smooth muscle
Pulmonary artery
Pulmonary vein
Respiratory bronchioles
Capillary beds
Alveoli
Alveolar duct
Alveolar pore
Alveolar sac

Alveoli

The respiratory bronchioles eventually become alveolar ducts, which terminate in clusters of alveoli surrounded by capillaries (*alveolar sacs*). Gas exchange takes place through the alveoli.

Alveolar walls contain two basic epithelial cell types:

Type I cells are the most abundant. It is across these thin, flat, squamous cells that gas exchange occurs.

Type II cells secrete *surfactant*, a substance that coats the alveolus and reduces surface tension. This allows the alveoli to remain inflated so that gas exchange can occur by diffusion. Surfactant is formed relatively late in foetal life; thus premature infants born without adequate amounts experience respiratory distress and may die.

Lungs

The cone-shaped lungs are located in the thoracic cavity and are surrounded by pleura. The right lung is shorter, broader and larger than the left. It has three lobes and handles 55% of gas exchange. The left lung has two lobes and contains a space for the heart (*cardiac notch*). Each lung's concave base rests on the diaphragm; the apex extends about 1.5 cm above the first rib.

> We hang out in the thoracic cavity and are surrounded by pleural membranes.

Pleura and pleural cavities

The pleura—the membrane that totally encloses the lung—is composed of a visceral layer and a parietal layer. The visceral pleura covers the entire lung surface, including the areas between the lobes. The parietal pleura lines the inner surface of the chest wall and upper surface of the diaphragm.

Serous fluid has serious functions

The pleural cavity—a potential space between the visceral and parietal pleural layers—contains a thin film of serous fluid. This fluid has two functions:

It lubricates the pleural surfaces, which allows them to slide smoothly against each other as the lungs expand and contract.

It creates a bond between the layers that causes the lungs to move with the chest wall during the mechanical breathing process.

Thoracic cavity

The thoracic cavity is surrounded by the ribs and muscles of the chest, the sternum and the thoracic vertebrae. The diaphragm forms the lower part of the thoracic cavity, separating the thoracic from the abdominal cavity.

Mediastinum

The space between the lungs is called the *mediastinum*. It contains the:
* heart and pericardium
* thoracic aorta

- pulmonary artery and veins
- venae cavae
- thymus, lymph nodes and vessels
- trachea, oesophagus and thoracic duct (part of the lymphatic system)
- vagus, cardiac and phrenic nerves.

Thoracic cage

The thoracic cage is composed of bone and cartilage. It supports and with the muscles protects the lungs, allowing them to expand and contract.

Posterior thoracic cage
The vertebral column and 12 pairs of ribs form the posterior portion of the thoracic cage. The ribs form the major portion of the thoracic cage. They extend from the thoracic vertebrae towards the anterior thorax.

Anterior thoracic cage
The anterior thoracic cage consists of the sternum (comprising the *manubrium* and *xiphoid process*) and ribs. The sternum protects the structures that lie between the lungs.

Attached or floating free?

Ribs pairs 1 through 7 attach directly to the sternum; pairs 8 through 10 attach to the cartilage of the preceding rib. The other 2 pairs of ribs are 'free-floating'—they don't attach to any part of the anterior thoracic cage. The ribs of pair 11 end anterolaterally, and pair 12 ends laterally.

Bordering on the costal angle

The lower parts of the rib cage (costal margins) near the xiphoid process form the borders of the costal angle—an angle of about 90 degrees in a normal person. (See *Locating lung structures in the thoracic cage*, page 156.)

It's suprasternal

Above the anterior thorax is a depression called the suprasternal notch. This dip at the base of the neck is sometimes assessed during physical examinations in which a patient is checked for signs of obvious health problems. Palpation of the suprasternal notch should not reveal a palpable pulse, except in some older patients. A prominent pulse in this area may indicate an aortic arch aneurysm.

The ribs, like the vertebrae, are numbered from top to bottom.

Because the suprasternal notch isn't covered by the rib cage, tracheal and aortic pulsation can be palpated here.

Body shop

Locating lung structures in the thoracic cage

The ribs, vertebrae and other structures of the thoracic cage act as landmarks that can be used to identify underlying structures as part of physical health assessment exercises.

From an anterior view

- The base of each lung rests at the level of the sixth rib at the midclavicular line and the eighth rib at the midaxillary line.
- The apex of each lung extends about 2 to 4 cm above the inner aspects of the clavicles.
- The sternal angle (angle of Louis) is a joint that lies between the manubrium and the main part of the sternum.
- The upper lobe of the right lung ends level with the fourth rib at the midclavicular line and with the fifth rib at the midaxillary line.
- The middle lobe of the right lung extends triangularly from the fourth to the sixth rib at the midclavicular line and to the fifth rib at the midaxillary line.
- Because the left lung doesn't have a middle lobe, the upper lobe of the left lung ends level with the fourth rib at the midclavicular line and with the fifth rib at the midaxillary line.

From a posterior view

- The lungs extend from the cervical area to the level of T10. On deep inspiration, the lungs may descend to T12.
- An imaginary line, stretching from the T3 level along the inferior border of the scapulae to the fifth rib at the midaxillary line, separates the upper lobes of both lungs.
- The upper lobes are situated above T3; the lower lobes are situated below T3 and extend to the level of T10.
- The diaphragm originates around the ninth or tenth rib.

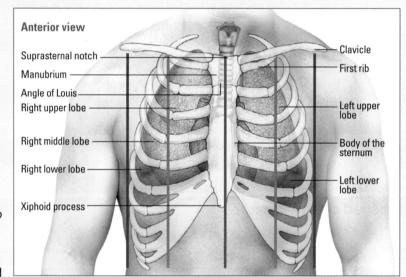

Anterior view

- Suprasternal notch
- Manubrium
- Angle of Louis
- Right upper lobe
- Right middle lobe
- Right lower lobe
- Xiphoid process
- Clavicle
- First rib
- Left upper lobe
- Body of the sternum
- Left lower lobe

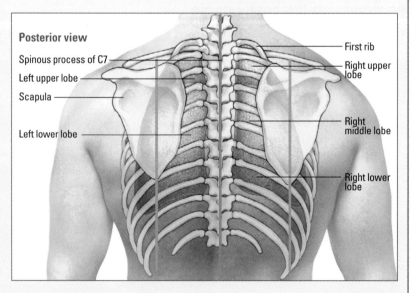

Posterior view

- Spinous process of C7
- Left upper lobe
- Scapula
- Left lower lobe
- First rib
- Right upper lobe
- Right middle lobe
- Right lower lobe

Inspiration and expiration

Breathing involves two actions: inspiration (an active process) and expiration (usually a passive process). Both actions rely on respiratory muscle function and the effects of pressure differences in the lungs.

Breathing involves two actions: inspiration and expiration.

It's perfectly normal!

During normal breathing, the external intercostal muscles aid the diaphragm, the major muscle of involved in ventilation. The dome-shaped diaphragm descends to lengthen the chest cavity, while the external intercostal muscles (located between and along the lower borders of the ribs) contract to expand the anteroposterior diameter. This coordinated action causes a reduction in intrapleural pressure, and inspiration occurs. Rising of the diaphragm and relaxation of the intercostal muscles causes an increase in intrapleural pressure, and expiration results. (See *Muscles of respiration*, page 158.)

Forced inspiration and active expiration

During exercise, when the body needs increased oxygenation, or in certain disease states that require forced inspiration and active expiration, the accessory muscles of respiration also participate.

When the body's demand for oxygen is increased, such as during exercise, the accessory muscles of respiration assist in the ventilation process.

Forced inspiration
During forced inspiration:
• the pectoral muscles in the upper chest raise the chest to increase the anteroposterior diameter
• the sternocleidomastoid muscles in the side of the neck raise the sternum
• the scalene muscles in the neck elevate, fix and expand the upper chest
• the posterior trapezius muscles in the upper back raise the thoracic cage.

Active expiration
During active expiration, the internal intercostal muscles contract to shorten the chest's transverse diameter and the abdominal rectus muscles pull down the lower chest, thus depressing the lower ribs. (See *Mechanics of ventilation*, page 159.)

(Text continues on page 160.)

Body shop

Muscles of respiration

The muscles of respiration help the chest cavity expand and contract. Pressure differences between atmospheric air and the lungs help produce air movement. These illustrations show the muscles that work together to allow inspiration and expiration.

Anterior view

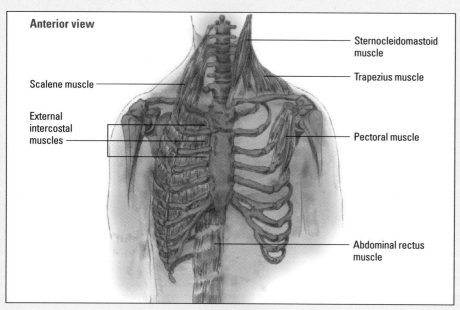

Sternocleidomastoid muscle

Scalene muscle

Trapezius muscle

External intercostal muscles

Pectoral muscle

Abdominal rectus muscle

Posterior view

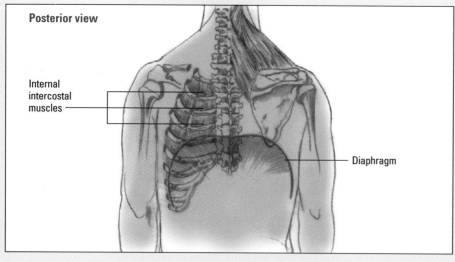

Internal intercostal muscles

Diaphragm

Now I get it!

Mechanics of ventilation

Breathing results from differences between atmospheric and intrapulmonary pressures, as described below.

Before inspiration, intrapulmonary pressure equals atmospheric pressure (approximately 101.3 kPa). Intrapleural pressure is 100.8 kPa.

During normal expiration, the diaphragm slowly relaxes and the lungs and thorax passively return to resting size and position. During deep or forced expiration, contraction of internal intercostal and abdominal muscles reduces thoracic volume. Lung and thorax compression raises intrapulmonary pressure above atmospheric pressure.

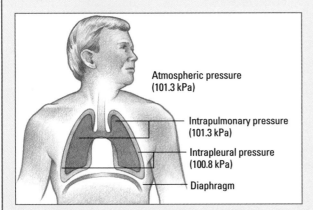

Atmospheric pressure
(101.3 kPa)

Intrapulmonary pressure
(101.3 kPa)

Intrapleural pressure
(100.8 kPa)

Diaphragm

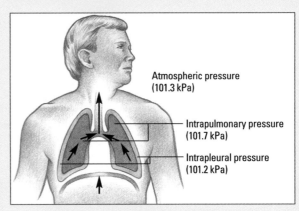

Atmospheric pressure
(101.3 kPa)

Intrapulmonary pressure
(101.7 kPa)

Intrapleural pressure
(101.2 kPa)

During inspiration, the diaphragm and external intercostal muscles contract, enlarging the thorax vertically and horizontally. As the thorax expands, intrapleural pressure decreases and the lungs expand to fill the enlarging thoracic cavity.

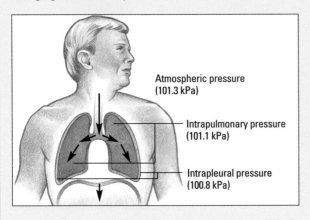

Atmospheric pressure
(101.3 kPa)

Intrapulmonary pressure
(101.1 kPa)

Intrapleural pressure
(100.8 kPa)

Fair exchange

De-oxygenated blood is pumped from the right ventricle into the pulmonary trunk, which divides into the right and left pulmonary arteries, carrying de-oxygenated blood to the lungs. The arteries become smaller arterioles and finally form the capillaries surrounding the alveoli. Gas exchange—oxygen and carbon dioxide diffusion—takes place in the alveoli.

Internal and external respiration

Effective respiration consists of gas exchange in the lungs, called *external respiration*, and gas exchange in the tissues, called *internal respiration*.

Internal respiration occurs only through diffusion. External respiration occurs through three processes:

ventilation—the process of breathing; getting air into and out of the pulmonary airways

pulmonary perfusion—blood flow from the right side of the heart, through the pulmonary circulation and into the left side of the heart

diffusion—gas movement through a selectively permeable membrane from an area of greater concentration to one of lesser concentration.

Ventilation

Ventilation is the breathing of gases (oxygen and carbon dioxide) into and out of the pulmonary airways. Problems within the nervous, musculoskeletal and pulmonary systems greatly compromise breathing effectiveness. (See *Respiratory changes with aging*.)

The central nervous system's respiratory centre is located in the pons with input from the medulla oblongata.

Nervous system influence

Involuntary breathing results from stimulation of the respiratory centres in the pons and medulla oblongata of the brain. Central chemical receptors in the medulla oblongata indirectly monitor the level of carbon dioxide in the blood. Carbon dioxide exerts the main influence on breathing. When carbon dioxide levels rise, the rate and depth of breathing increases to eliminate excess carbon dioxide.

Peripheral chemical receptors in the aorta and carotid arteries monitor the level of oxygen in the blood. When oxygen levels drop, respiratory rate and depth increase to improve the blood oxygen level. However, the peripheral chemical receptors are less sensitive than the central receptors and don't respond until oxygen levels are quite low.

As time goes by . . .

Respiratory changes with aging

As a person ages, the body undergoes respiratory system changes. These changes can include structural changes as well as changes in function.

Structural changes

Age-related anatomical changes in the upper airways include nose enlargement from continued cartilage growth, general atrophy of the tonsils and tracheal deviations from changes in the aging spine. Possible thoracic changes include increased anteroposterior chest diameter (resulting from altered calcium metabolism) and calcification of costal cartilage, which reduce mobility of the chest wall. Kyphosis advances with age because of such factors as osteoporosis and vertebral collapse.

The lungs become more rigid and the number and size of alveoli decline with age. In addition, a 30% reduction in respiratory fluids heightens the risk of pulmonary infection and mucus plugs.

Pulmonary function changes

Pulmonary function decreases in older people as a result of respiratory muscle degeneration or atrophy. Ventilatory capacity diminishes for several reasons:

- The lungs' diffusing capacity declines; decreased inspiratory and expiratory muscle strength diminishes vital capacity.
- Lung tissue degeneration causes a decrease in the lungs' elastic recoil capability, which results in an elevated residual volume. Thus, aging alone can cause emphysema.
- Closing of some airways produces poor ventilation of the basal areas, resulting in both a decreased surface area for gas exchange and reduced partial pressure of oxygen. The partial pressure of oxygen in arterial blood may decrease to around 11.5 kPa, and oxygen saturation reduces by approximately 5%.

Musculoskeletal influence

The adult thorax is flexible—its shape can be changed by contracting the chest muscles. The medulla oblongata controls ventilation primarily by stimulating contraction of the diaphragm and external intercostal muscles. These actions cause the lungs to expand, producing the intrapulmonary pressure changes that cause inspiration.

Pulmonary influence

Airflow distribution can be affected by many factors:
- airflow pattern (see *Comparing airflow patterns*, page 162.)
- volume and location of the functional reserve capacity (air retained in the alveoli that prevents their collapse during expiration)
- degree of intrapulmonary resistance
- presence of lung disease

Comparing airflow patterns

The pattern of airflow through the respiratory passages affects airway resistance.

Laminar flow

Laminar flow, a linear pattern that occurs at low flow rates, offers minimal resistance.

This flow type occurs mainly in the small peripheral airways of the bronchial tree.

Turbulent flow

The eddying pattern of turbulent flow creates friction and increases resistance.

Turbulent flow is normal in the trachea and large central bronchi. If the smaller airways become constricted or clogged with secretions, however, turbulent flow may also occur there.

Transitional flow

A mixed pattern known as transitional flow is common at

lower flow rates in the larger airways, especially where the airways narrow from obstruction, meet or branch.

The path of least resistance

If airflow is disrupted for any reason, airflow distribution follows the path of least resistance.

Increased workload, decreased efficiency

Other musculoskeletal and intrapulmonary factors can affect airflow and, in turn, may affect the breathing pattern and *tidal volume* (the volume of air that is inhaled and exhaled during the normal breathing process). For instance, forced breathing (as occurs in emphysema) activates accessory muscles of respiration, which require additional oxygen to work. This results in less efficient ventilation with an increased workload on the body.

Airflow interference and alterations

Other airflow alterations can also increase oxygen and energy demand and cause respiratory muscle fatigue. These conditions include interference with expansion of the lungs or thorax (changes in compliance) and interference with airflow in the tracheobronchial tree (changes in resistance). Both can result in reduced tidal volume and alveolar ventilation.

Pulmonary perfusion

Pulmonary perfusion refers to blood flow from the right side of the heart, through the pulmonary circulation, and into the left side of the heart. Perfusion aids external respiration. Normal pulmonary blood flow allows alveolar gas exchange, but many factors may interfere with gas transport to the alveoli. Here are some examples:

• Cardiac output less than the average of 5 L/minute decreases gas exchange by reducing blood flow.
• Elevations in pulmonary and systemic resistance reduce blood flow.
• Abnormal or insufficient haemoglobin picks up less oxygen for exchange.

Ventilation-perfusion match

Efficient gas exchange requires that the amount of gases in the alveoli (ventilation) matches the blood flow through the capillaries (perfusion). (See *What happens in ventilation-perfusion mismatch*, page 164.)

Diffusion

In diffusion, oxygen and carbon dioxide molecules move between the alveoli and capillaries. The direction of movement is always from an area of greater concentration to one of lesser concentration. In the process, oxygen moves across the alveolar and capillary membranes, dissolves in the plasma, and then passes through the red blood cell (RBC) membrane. Carbon dioxide moves in the opposite direction.

The interesting thing about interstitial spaces

The epithelial membranes lining the alveoli and capillaries must be intact. Both the alveolar epithelium and the capillary endothelium are composed of a single layer of cells. Between these layers are tiny interstitial spaces filled with elastin and collagen. Thickening in the interstitial spaces can slow diffusion.

From the RBCs to the alveoli

Normally, oxygen and carbon dioxide move easily through all of these layers. Oxygen moves from the alveoli into the bloodstream, where it's taken up by haemoglobin in the RBCs. When oxygen arrives in the bloodstream, it displaces carbon dioxide (the by-product of metabolism), which diffuses from RBCs into the blood and then to the alveoli.

During diffusion, oxygen and carbon dioxide travel the same path but in opposite directions!

Move over! This is my place now.

Now I get it!

What happens in ventilation-perfusion mismatch

Ideally, the amount of air in the alveoli (a reflection of ventilation) matches the amount of blood in the capillaries (a reflection of perfusion). This allows gas exchange to proceed smoothly.

However, this ventilation-perfusion ($\dot{V}/\dot{Q}$) ratio is actually unequal: the alveoli receive air at a rate of approximately 4 L/minute, while the capillaries supply blood at a rate of about 5 L/minute. This creates a $\dot{V}/\dot{Q}$ mismatch of 4:5, or 0.8.

Normal

In the normal lung, ventilation closely matches perfusion.

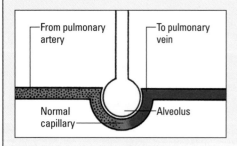

Dead-space ventilation

Normal ventilation without adequate perfusion usually results from a perfusion defect such as pulmonary embolism.

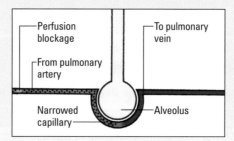

Shunt

Perfusion without adequate ventilation usually results from airway obstruction, particularly that caused by acute diseases, such as atelectasis and pneumonia.

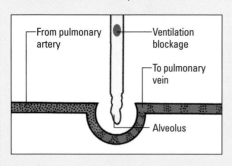

Silent unit

Inadequate ventilation and perfusion usually stems from multiple causes, such as pulmonary embolism with resultant acute respiratory distress syndrome and emphysema.

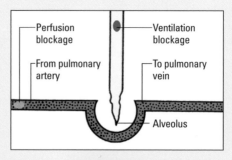

 Blood with CO_2 Blood with O_2 Blood with CO_2 and O_2

To bind or not to bind

Most transported oxygen binds with haemoglobin to form oxyhaemoglobin; however, a small portion dissolves in the plasma (approximately 2%). The portion of oxygen that dissolves in plasma can be measured as the partial pressure of oxygen in arterial blood (Po_2).

After oxygen binds to haemoglobin, RBCs travel to the tissues. Through cellular diffusion, internal respiration occurs when RBCs release oxygen. Carbon dioxide then diffuses into RBCs and plasma for transport back to the lungs for removal during expiration. Roughly 23% of the carbon dioxide carried in blood is found bound to haemoglobin. However, the main method of carbon dioxide transport in blood is as bicarbonateion (approximately 70% is carried in this way), with the remainder dissolved in plasma.

Acid-base balance

Oxygen taken up by the lungs is transported to the tissues by the circulatory system (bound to haemoglobin or dissolved in plasma). Carbon dioxide produced by cellular metabolism is carried back to the lungs dissolved in plasma as biocarbonate ion (base) or carbonic acid (acid). A small amount of carbon dioxide is carried bound to haemoglobin as carbaminohaemoglobin. The respiratory system uses hypoventilation or hyperventilation as needed to regulate acidity.

Respiratory responses

The lungs control bicarbonate levels by converting bicarbonate to carbon dioxide and water for excretion. In response to signals from the pons and medulla oblongata, the lungs can change the rate and depth of breathing. This change allows for adjustments in the amount of carbon dioxide lost, to help maintain acid-base balance.

Metabolic alkalosis

For example, in *metabolic alkalosis* (a condition resulting from excess bicarbonate retention), the rate and depth of ventilation decrease so that carbon dioxide can be retained; this increases carbonic acid levels.

Metabolic acidosis

In *metabolic acidosis* (a condition resulting from excess acid retention or excess bicarbonate loss), the lungs increase the rate and depth of ventilation to eliminate excess carbon dioxide, thus reducing carbonic acid levels.

The pons and medulla oblongata change the rate and depth of breathing. That affects how much CO_2 we get rid of!

Imbalance woes

When the lungs don't function properly, an acid-base imbalance results. For example, they can cause respiratory acidosis through *hypoventilation* (reduced rate and depth of alveolar ventilation), which leads to carbon dioxide retention. Conversely, respiratory alkalosis results from *hyperventilation* (increased rate and depth of alveolar ventilation), which leads to carbon dioxide elimination.

Quick quiz

1. Which of the following respiratory structures is the main site of gas exchange?
    A. Alveoli
    B. Terminal bronchioles
    C. Respiratory bronchioles
    D. Pulmonary arteries

Answer: A. The alveoli is the main site for gas exchange within the lungs.

2. How many lobes does the right lung have?
    A. Six
    B. Two
    C. Three
    D. One

Answer: C. The right lung has three lobes.

3. During external gas respiration, oxygen and carbon dioxide diffusion occurs in the:
    A. Venules.
    B. Alveoli.
    C. Red blood cells.
    D. Body tissues.

Answer: B. Oxygen and carbon dioxide diffusion occurs in the alveoli.

4. When oxygen passes through the alveoli into the bloodstream, it binds with haemoglobin to form:
    A. Red blood cells.
    B. Carbon dioxide.
    C. Nitrogen.
    D. Oxyhaemoglobin.

Answer: D. When oxygen passes through the alveoli into the bloodstream, it binds with haemoglobin to form oxyhaemoglobin.

Scoring

☆☆☆   If you answered all four questions correctly, extraordinary! Take a deep breath! You've responded extremely well to the respiratory system.

☆☆   If you answered three questions correctly, fascinating! You're breezing through these systems like a whirlwind.

☆   If you answered fewer than three questions correctly, get inspired! Perhaps you'll catch your second wind the next time through the chapter.

Thanks for studying hard. We can breathe much easier now.

Just the facts

In this chapter, you'll learn:

♦ two major components of the gastrointestinal system

♦ phases of digestion

♦ functions of gastrointestinal hormones

♦ sites and mechanisms of gastric secretions.

A look at the GI system

The gastrointestinal (GI) system has two major components: the *alimentary canal* (also called the *GI tract*) and the *accessory GI organs*. The GI tract serves two major functions:

🖐 *digestion*, or the breaking down of food and fluid into simple chemicals that can be absorbed into the bloodstream and transported throughout the body

✌ *elimination* of waste products through excretion of faeces

Alimentary canal

The alimentary canal is a hollow muscular tube that begins in the mouth and extends to the anus. It includes the mouth and tongue, pharynx, oesophagus, stomach, small intestine and large intestine. (See *Structures of the GI system*.) The wall of the alimentary canal is made up of four layers.

What goes in must come out. Digestion and excretion are the GI tract's major functions.

Body shop

Structures of the GI system

The GI system includes the alimentary canal (pharynx, oesophagus, stomach and small and large intestines) and the accessory organs (liver, biliary duct system and pancreas).

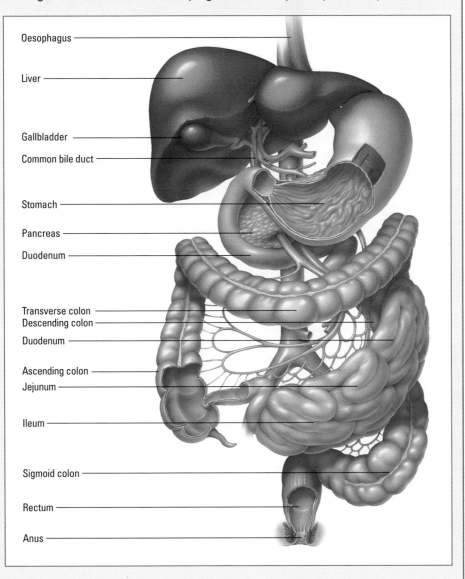

Oesophagus

Liver

Gallbladder

Common bile duct

Stomach

Pancreas

Duodenum

Transverse colon

Descending colon

Duodenum

Ascending colon

Jejunum

Ileum

Sigmoid colon

Rectum

Anus

Mouth

The mouth (also called the *buccal cavity* or *oral cavity*) is bounded by the lips (labia), cheeks, palate (roof of the mouth) and tongue, and contains the teeth. Ducts connect the mouth with the three major pairs of salivary glands (parotid, submandibular, sublingual). These glands secrete *saliva* which acts to moisten food and helps to a food *bolus*, so it can be swallowed easily.

Saliva contains the enzyme amylase which assists in the breakdown of carbohydrates. Thus, digestion of food begins in the mouth. Salivary glands also secrete salivary lipase (a more potent form of lipase) to start fat digestion. This form of lipase plays a large role in fat digestion in newborns as their pancreatic lipase still has some time to develop.

Saliva also has a protective function, helping to prevent bacterial build-up on the teeth and washing away adhered food particles. The mouth initiates the mechanical breakdown of food. (See *Oral cavity*.)

Moisten, chew and break us down.

Pharynx

The *pharynx* is a cavity that extends from the base of the skull to the oesophagus. The muscular pharynx aids swallowing by propelling food towards the oesophagus. When food enters the pharynx, the *epiglottis* (a flap of connective tissue) closes over the trachea to prevent aspiration.

Oesophagus

The *oesophagus* is a muscular tube that extends from the pharynx through the mediastinum to the stomach. Swallowing triggers the passage of food from the pharynx to the oesophagus. The cricopharyngeal sphincter—a sphincter at the upper border of the oesophagus—must relax for food to enter the oesophagus. Peristaltic waves propel liquids and solids through the oesophagus into the stomach.

Stomach

The *stomach* is a distensible enlargement of the GI tract in the left upper portion of the abdominal cavity, just below the diaphragm. Its upper border attaches to the lower end of the oesophagus. The lateral surface of the stomach is called the *greater curvature*; the medial surface, the *lesser curvature*.

Size does matter!

The size of the stomach varies with the degree of distention. Overeating can cause marked distention, which pushes on the diaphragm and causes shortness of breath.

Body shop

Oral cavity

The mouth, or oral cavity, is bounded by the lips (labia), cheeks, palate (roof of the mouth) and tongue. The mouth initiates the mechanical breakdown of food.

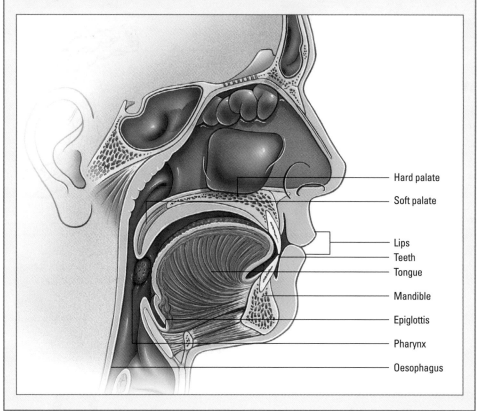

Hard palate
Soft palate
Lips
Teeth
Tongue
Mandible
Epiglottis
Pharynx
Oesophagus

The Fab four

The stomach has four main regions:

The *cardia* lies near the junction of the stomach and oesophagus.

The *fundus* is an enlarged portion above and to the left of the oesophageal opening into the stomach.

The *body* is the middle portion of the stomach.

The *pylorus* is the lower portion, lying near the junction of the stomach and duodenum.

Just passing through

The stomach has several functions, including:
- serving as a temporary storage area for food
- beginning of protein digestion and continuation of carbohydrate breakdown
- breaking down food into *chyme*, a semifluid substance
- release of the gastric content into the small intestine via the pyloric sphincter.

Small intestine

The *small intestine* is a tube that measures about 6 m in length. It's the longest organ of the GI tract and has three major divisions:
- The *duodenum* is the shortest and most superior division.
- The *jejunum* is the middle portion.
- The *ileum* is the longest and most inferior portion. (See *A look at special cells*.)

Intestinal wall
The intestinal wall has structural features that significantly increase its absorptive surface area. These features include *plicae circulares*—circular folds of the intestinal mucosa, or mucous membrane lining.

Free villi

Villi and microvilli are also intestinal wall features that increase the absorptive area of the intestinal wall. *Villi* are fingerlike projections on the mucosa. *Microvilli* are tiny cytoplasmic projections on the surface of epithelial cells.

Other structures
The small intestine also contains intestinal crypts, Peyer's patches and Brunner's glands:
- *Intestinal crypts* are simple glands lodged in the grooves separating villi.
- *Peyer's patches* are collections of lymphatic tissue within the submucosa.
- *Brunner's glands* secrete mucus.

Functions
Functions of the small intestine include:
- completing food digestion
- absorbing food molecules through its wall into the circulatory system, which then delivers them to body tissues and cells
- secreting hormones that help control the secretion of bile, pancreatic fluid and intestinal fluid.

The small intestine is a tube that measures about 6 m in length. It's the longest organ of the GI tract.

Completing digestion, absorbing food and controlling the secretion of bile...the small intestine has a lot of work to do!

Zoom in

A look at special cells

This cross section of the stomach shows the G-cells (which secrete gastrin) in the pyloric glands. The cross section of the small intestine show the S-cells (which produce secretin) in the duodenal and jejunal glands.

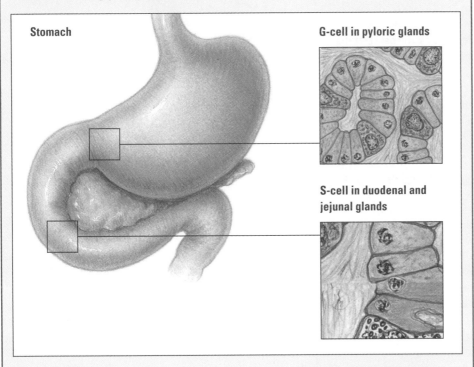

Stomach

G-cell in pyloric glands

S-cell in duodenal and jejunal glands

Large intestine

The *large intestine* extends from the ileocaecal valve (the valve between the ileum of the small intestine and the first segment of the large intestine) to the anus. It has six segments:

The *caecum*, a saclike structure, makes up the first few centimetres.

The *ascending colon* rises on the right posterior abdominal wall, then turns sharply under the liver at the hepatic flexure.

The *transverse colon* is situated above the small intestine, passing horizontally across the abdomen and below the liver, stomach and spleen. At the left colic flexure, also known as the *splenic flexure*, it turns downward.

The *descending colon* starts near the spleen and extends down the left side of the abdomen into the pelvic cavity.

The *sigmoid colon* descends through the pelvic cavity, where it becomes the rectum.

The *rectum*, the last few centimetres of the large intestine, terminates at the *anus*, which is the external opening of the large intestine that allows expulsion of waste products.

Functions

The functions of the large intestine include absorbing water, secreting mucus and eliminating digestive wastes. It also contains large numbers of bacteria which produce some vitamins.

GI tract wall structures

The wall of the GI tract consists of several layers. These layers are the mucosa, submucosa, muscularis, and visceral peritoneum.

Mucosa

The *mucosa*, the innermost layer, consists of epithelial and surface cells and loose connective tissue. *Villi*, fingerlike projections of the mucosa, secrete gastric and protective juices and absorb nutrients.

Submucosa

The *submucosa* encircles the mucosa. It's composed of loose connective tissue, blood and lymphatic vessels and a nerve network called the *submucosal plexus*, or *Meissner's plexus*.

Muscularis

The *muscularis*, which lies around the submucosa, is composed of skeletal muscle in the mouth, pharynx and upper oesophagus.

Fibres, fibres everywhere

Elsewhere in the tract, the muscularis is made up of longitudinal and circular smooth muscle fibres. During peristalsis, the involuntary wave-like movement of the intestines, longitudinal fibres shorten the lumen length and circular fibres reduce the lumen diameter. At points along the tract, circular fibres thicken to form sphincters.

Past the lips, past the gums...look out, mucosa, here it comes!

All pouched up

In the large intestine, these fibres gather into three narrow bands (*taeniae coli*) down the middle of the colon and pucker the intestine into characteristic pouches (*haustra*).

Networking is key

Between the two muscle layers lies another nerve network–the *myenteric plexus*, also known as *Auerbach's plexus*. The stomach wall contains a third muscle layer made up of oblique fibres. (See *Features of the GI tract wall*, page 176.)

Visceral peritoneum

The *visceral peritoneum* (also known as the *tunica serosa*) is the GI tract's outer covering. It covers most of the abdominal organs and lies next to an identical layer, the *parietal peritoneum*, which lines the abdominal cavity.

A double-layered fold

The visceral peritoneum becomes a double-layered fold around the blood vessels, nerves and lymphatics. It attaches the jejunum and ileum to the posterior abdominal wall to prevent twisting. A similar fold attaches the transverse colon to the posterior abdominal wall.

> So many ways of saying one thing! The visceral peritoneum is known as the tunica adventitia in one part of the body and the tunica serosa in another.

A visceral peritoneum by any other name

The visceral peritoneum has many names. In the oesophagus and rectum, it's called the *tunica adventitia*; elsewhere in the GI tract, it's called the *tunica serosa*.

GI tract innervation

Distention of the submucosal plexus stimulates transmission of nerve signals to the smooth muscle, which initiates peristalsis and mixing contractions.

Parasympathetic stimulation

Parasympathetic stimulation of the vagus nerve (for most of the intestines) and the sacral spinal nerves (for the descending colon and rectum) increases gut and sphincter tone. It also increases the frequency, strength and velocity of smooth-muscle contractions as well as motor and secretory activities.

Zoom in

Features of the GI tract wall

Four layers—the mucosa, submucosa, muscularis and visceral peritoneum—form the wall of the GI tract. This illustration depicts the cellular anatomy of the jejunum. Many of the obvious anatomical features of the mucosa function to increase the absorptive area of the structure.

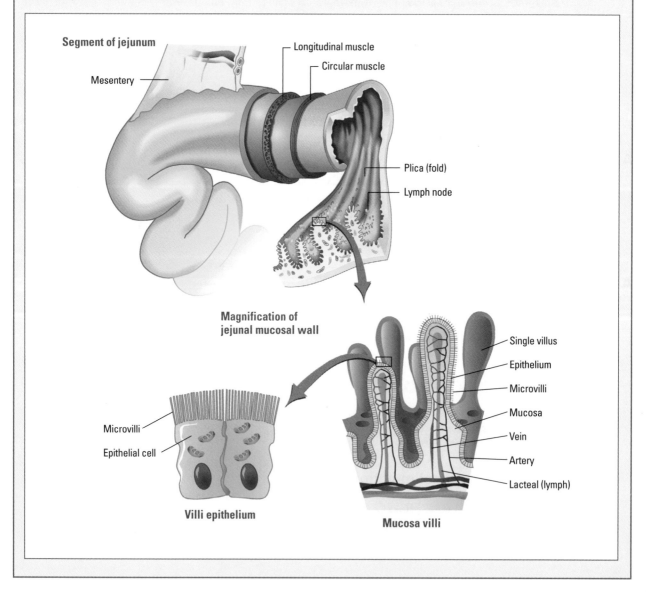

Segment of jejunum

Mesentery

Longitudinal muscle

Circular muscle

Plica (fold)

Lymph node

Magnification of jejunal mucosal wall

Single villus

Epithelium

Microvilli

Mucosa

Vein

Artery

Lacteal (lymph)

Microvilli

Epithelial cell

Villi epithelium

Mucosa villi

Sympathetic stimulation

Sympathetic stimulation, by way of the spinal nerves from levels T6 to L2, reduces peristalsis and inhibits GI activity.

Accessory GI organs and glands

A little respect please! I'm the largest gland inside the body.

The accessory GI organs and glands—the liver, gallbladder, pancreas and salivary glands—contribute hormones, enzymes and bile to the digestive process.

Liver

The body's largest gland, the 1.4 kg, highly vascular liver is enclosed in a fibrous capsule in the right upper quadrant of the abdomen. The *lesser omentum*, a fold of peritoneum, covers most of the liver and anchors it to the lesser curvature of the stomach. The *hepatic artery* and *hepatic portal vein*, as well as the common bile duct and hepatic veins, pass through the lesser omentum.

Lobes and lobules

The liver consists of four lobes:
- left lobe
- right lobe
- caudate lobe (behind the right lobe)
- quadrate lobe (behind the left lobe).

Function . . .

The liver's functional unit, the *lobule*, consists of a plate of hepatic cells, or *hepatocytes*, that encircle a central vein and radiate outward. Separating the hepatocyte plates from each other are *sinusoids*, the liver's capillary system. Reticuloendothelial macrophages (Kupffer cells) that line the sinusoids remove bacteria and toxins that have entered the blood through the intestinal capillaries. (See *A look at a liver lobule*, page 178.)

. . . and flow

The sinusoids carry oxygenated blood from the hepatic artery and nutrient-rich blood from the portal vein. Deoxygenated blood leaves through the central vein and flows through hepatic veins to the inferior vena cava.

Zoom in

A look at a liver lobule

The liver's functional unit is called a *lobule*. It consists of a plate of hepatic cells, or hepatocytes, that encircle a central vein and radiate outward. Separating the hepatocyte plates from each other are *sinusoids*, which serve as the liver's capillary system. Sinusoids carry oxygenated blood from the hepatic artery and nutrient-rich blood from the portal vein.

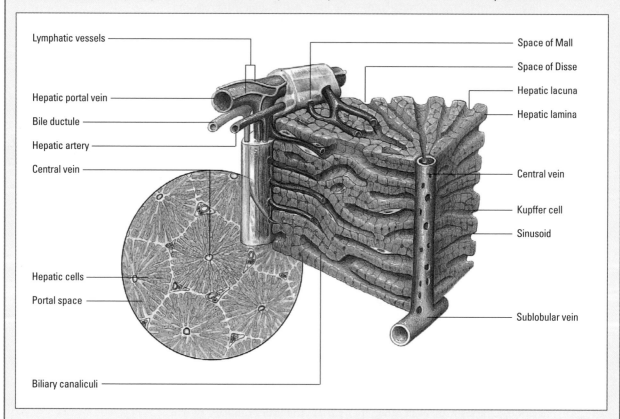

Ducts

The ductile system is composed of a series of interconnected tubes that transport bile through the GI tract. *Bile* is a greenish liquid composed of water, cholesterol, bile salts and phospholipids. It exits through bile ducts (canaliculi) that merge into the right and left hepatic ducts to form the common hepatic duct. This duct joins the cystic duct from the gallbladder to form the common bile duct that leads to the duodenum.

Job description

The liver serves several important functions:
• It plays an important role in carbohydrate metabolism including the storage of glucose in the form of glycogen.
• It detoxifies various endogenous and exogenous toxins in plasma.
• It synthesises plasma proteins, nonessential amino acids and stores vitamin A.
• It stores essential nutrients, such as vitamins K, D and B$_{12}$, and iron.
• It removes ammonia from body fluids, converting it to urea for excretion in urine.
• It helps regulate blood glucose levels.
• It secretes bile.

Function of bile

Bile has several functions, including:
• emulsifying (breaking down) fat
• promoting intestinal absorption of fatty acids, cholesterol and other lipids.

When bile salts are missing

When bile salts are absent from the intestinal tract, lipids are excreted and fat-soluble vitamins are poorly absorbed.

Report on bile production

The liver recycles about 80% of bile salts into bile, combining them with bile pigments (biliverdin and bilirubin, the waste products of red blood cell breakdown) and cholesterol. The liver continuously secretes this alkaline bile. Bile production may increase from stimulation of the vagus nerves, release of the hormone secretin, increased blood flow in the liver and the presence of fat in the intestine. (See *GI hormones: production and function*, page 180.)

Gallbladder

The *gallbladder* is a pear-shaped organ joined to the inferomedial surface of the liver by the cystic duct. It's covered with visceral peritoneum.

On the job

The gallbladder stores and concentrates bile produced by the liver. It also releases bile into the common bile duct (formed by the cystic duct and common hepatic duct) for delivery to the duodenum in response to the contraction and relaxation of the hepatopancreatic sphincter (sphincter of Oddi).

My role in carbohydrate metabolism is very important. I also detoxify toxins in plasma.

Now I get it!

GI hormones: production and function

When stimulated, GI structures secrete four hormones. Each hormone plays a different role in digestion.

Hormone and production site	Stimulating factor or agent	Function
Gastrin Produced in pyloric antrum and duodenal mucosa	• Pyloric antrum distention • Vagal stimulation • Protein digestion products • Alcohol	Stimulates gastric secretion and motility
Gastric inhibitory peptide Produced in duodenal and jejunal mucosa	• Gastric acid • Fats • Fat digestion products	Inhibits gastric secretion and motility
Secretin Produced in duodenal and jejunal mucosa	• Gastric acid • Fat digestion products • Protein digestion products	Stimulates secretion of bile and alkaline pancreatic fluid
Cholecystokinin Produced in duodenal and jejunal mucosa	• Fat digestion products • Protein digestion products	Stimulates gallbladder contraction and secretion of enzyme-rich pancreatic fluid

Pancreas

The *pancreas* is a somewhat flattened organ that lies inferior to the stomach. Its head and neck extend into the curve of the duodenum and its tail lies against the spleen. (See *A look at the biliary tract*.) It performs both exocrine and endocrine functions.

Exocrine function

The pancreas's exocrine function involves scattered cells that secrete more than 1 L of digestive enzymes and alkaline pancreatic juice every day. Lobules and lobes of the clusters (*acini*) of enzyme-producing cells release their secretions into ducts that merge into the pancreatic duct.

Endocrine function

The endocrine function of the pancreas is performed by the islets of Langerhans, which are located between the acinar cells.

Memory jogger

To remember the difference between *exocrine* and *endocrine*, keep in mind that **ex**ocrine refers to **ex**ternal, and **endo**crine refers to **in**ternal.

Body shop

A look at the biliary tract

Together, the gallbladder and pancreas constitute the biliary tract. The structures of the biliary tract are depicted in the illustration below.

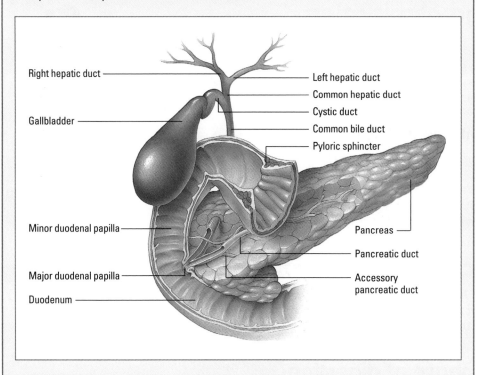

Right hepatic duct

Left hepatic duct

Common hepatic duct

Cystic duct

Gallbladder

Common bile duct

Pyloric sphincter

Minor duodenal papilla

Pancreas

Pancreatic duct

Major duodenal papilla

Accessory pancreatic duct

Duodenum

Alpha, beta . . . but no omega

The pancreatic islets are composed of four different types of cell which each produce different hormones. There are two main types of islet cells: alpha and beta. More than 1 million islets house these two cell types. Beta cells secrete *insulin* to promote carbohydrate metabolism; alpha cells secrete *glucagon*, a hormone that stimulates glycogenolysis breakdown of glycogen to glucose) in the liver. Both hormones flow directly into the blood. Their release is stimulated by blood glucose levels.

Pancreatic duct

Running the length of the pancreas, the *pancreatic duct* joins the bile duct from the gallbladder before entering the duodenum. Vagal stimulation and release of the hormones secretin and cholecystokinin control the rate and amount of pancreatic secretion.

Digestion and elimination

Digestion starts in the oral cavity, where chewing (*mastication*), salivation (the beginning of starch digestion) and swallowing (*deglutition*) take place. When a person swallows, the sphincter in the upper oesophagus relaxes, allowing food to enter the oesophagus. (See *What happens in swallowing*.)

Long day's journey into the stomach

In the oesophagus, the motor division of glossopharyngeal nerve innervates the pharynx for peristalsis and stimulates the salivary glands to release saliva. As food passes through the oesophagus, glands in the oesophageal mucosal layer secrete mucus, which lubricates the bolus and protects the mucosal membrane from damage caused by poorly chewed foods.

Cephalic phase of digestion

By the time the food bolus is travelling towards the stomach, the *cephalic* phase of digestion has already begun. In this phase, the stomach secretes digestive juices (hydrochloric acid [HCl] and pepsin).

Gastric phase of digestion

When food enters the stomach through the cardiac (lower oesophageal) sphincter, the stomach wall stretches, initiating the gastric phase of digestion. In this phase, distention of the stomach wall stimulates the stomach to release *gastrin*.

Gastrin

Gastrin stimulates the stomach's motor functions and secretion of gastric juice by the gastric glands. Highly acidic (pH of 1 to 3), these digestive secretions consist mainly of pepsin, HCl, intrinsic factor and proteolytic enzymes. (See *Sites and mechanisms of gastric secretion*, page 184.)

Chewing, salivation and swallowing. Let digestion begin!

Now I get it!

What happens in swallowing

Before peristalsis can begin, the neural pattern that initiates swallowing must occur. This process has a number of steps:
- Food pushed to the back of the mouth stimulates swallowing receptor areas that surround the pharyngeal opening.
- These receptor areas transmit impulses to the brain by way of the sensory portions of the trigeminal and glossopharyngeal nerves.
- The brain's swallowing centre then relays motor impulses to the oesophagus by way of the trigeminal, glossopharyngeal, vagus and hypoglossal nerves, causing swallowing to occur.

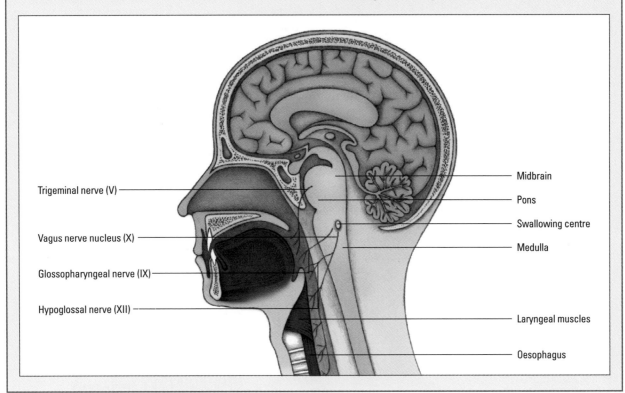

Trigeminal nerve (V)

Vagus nerve nucleus (X)

Glossopharyngeal nerve (IX)

Hypoglossal nerve (XII)

Midbrain

Pons

Swallowing centre

Medulla

Laryngeal muscles

Oesophagus

Intestinal phase of digestion

Normally, except for alcohol, minimal food absorption occurs in the stomach. Peristaltic contractions churn the food into tiny particles and mix it with gastric juices, forming *chyme*. Next, stronger peristaltic waves move the chyme into the antrum, where it backs up against the pyloric sphincter before being released into the duodenum, triggering the intestinal phase of digestion.

Now I get it!

Sites and mechanisms of gastric secretion

The body of the stomach lies between the cardiac and the pyloric sphincter. Between these sphincters lie the fundus, body, antrum and pylorus. These areas have a rich variety of mucosal cells that help the stomach carry out its tasks.

Glands and gastric secretions

Cardiac glands, pyloric glands and gastric glands secrete 2 to 3 L of gastric juice daily through the stomach's gastric pits. Here are the details:

- Both the *cardiac glands* (located in the region of the cardiac sphincter) and the *pyloric glands* (found near the pylorus) secrete thin mucus.
- The *gastric glands* (in the body and fundus) secretes hydrochloric acid (HCl), pepsinogen (the precursor of pepsin), intrinsic factor and mucus.

Protection from self-digestion

Specialised cells line the gastric glands, gastric pits and surface epithelium. Mucous cells in the necks of the gastric glands produce thin mucus. Mucous cells in the surface epithelium produce an alkaline mucus. Both substances lubricate food and protect the stomach from self-digestion by both acid and enzymes.

Other secretions

Gastrin is produced by G cells and acts to stimulate gastric secretion and motility. *Chief cells* produce pepsinogen which is converted to pepsin and breaks proteins down into polypeptides. Parietal cells scattered throughout the fundus secrete HCl and intrinsic factor. HCl degrades pepsinogen, maintains acid environment and inhibits excess bacteria growth. Intrinsic factor promotes vitamin B_{12} absorption in the small intestine.

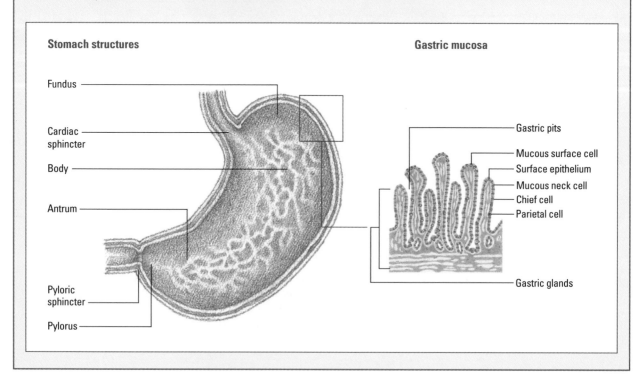

Stomach structures

Gastric mucosa

- Fundus
- Cardiac sphincter
- Body
- Antrum
- Pyloric sphincter
- Pylorus

- Gastric pits
- Mucous surface cell
- Surface epithelium
- Mucous neck cell
- Chief cell
- Parietal cell
- Gastric glands

Stomach emptying

The rate of stomach emptying depends on several factors, including gastrin release, neural signals generated when the stomach wall distends and the *enterogastric reflex*. In this reaction, the duodenum releases secretin and gastric-inhibiting peptide, and the jejunum secretes cholecystokinin—all of which act to decrease gastric motility.

Small intestine

The small intestine performs most of the work of digestion and absorption. (See *Small intestine: how form affects absorption*, page 186.)

Small but mighty

In the small intestine, intestinal contractions and various digestive secretions break down carbohydrates, proteins and fats—actions that enable the intestinal mucosa to absorb these nutrients into the bloodstream (along with water and electrolytes). These nutrients are then available for use by the body.

By the time chyme passes through the small intestine and enters the ascending colon of the large intestine, it has been reduced to mostly indigestible substances.

It only takes a whiff! Digestive juices are secreted in response to smelling, tasting, chewing or thinking about food.

Large intestine

The chyme begins its journey through the large intestine where the ileum and caecum join with the ileocaecal pouch. Then the chyme moves up the ascending colon and past the right abdominal cavity to the liver's lower border. It crosses in a roughly horizontal fashion below the liver and stomach, by way of the transverse colon, and descends the left abdominal cavity to the level of the iliac fossa through the descending colon.

Continuing journey of chyme

From there, the chyme travels through the sigmoid colon to the lower midline of the abdominal cavity, then to the rectum and finally to the anal canal. The anus opens to the exterior through two sphincters. The *internal sphincter* contains thick, circular smooth muscle under autonomic control; the *external sphincter* contains skeletal muscle under voluntary control.

Role in absorption

The large intestine produces no hormones or digestive enzymes; it continues the absorptive process. Through blood and lymph vessels in the submucosa, the proximal half of the large intestine absorbs all but about 100 ml of the remaining water in the colon. It also absorbs large amounts of sodium and chloride.

Now I get it!

Small intestine: how form affects absorption

Nearly all digestion and absorption takes place in the 6 m of small intestine. The structure of the small intestine, as shown below, is key to digestion and absorption.

Specialised mucosa

Multiple projections of the intestinal mucosa increase the surface area for absorption several hundredfold, as shown in the enlarged view at bottom left.

Circular folds are covered by villi. Each villus contains a lymphatic vessel (*lacteal*), a venule, capillaries, an arteriole, nerve fibres and smooth muscle.

Each villus is densely fringed with about 2,000 microvilli making it resemble a fine brush. The villi are lined with columnar epithelial cells, which dip into the lamina propria between the villi to form intestinal glands (*crypts of Lieberkühn*).

Types of epithelial cells

The type of epithelial cell dictates its function. Mucus-secreting goblet cells are found on and between the villi on the crypt mucosa. In the proximal duodenum, specialised Brunner's glands also secrete large amounts of mucus to lubricate and protect the duodenum from potentially corrosive acidic chyme and gastric juices.

Undifferentiated cells deep within the intestinal glands replace the epithelium. *Absorptive cells* consist of many tightly packed microvilli over a plasma membrane that contains transport mechanisms for absorption and produces enzymes for the final step in digestion.

Intestinal glands

The intestinal glands primarily secrete a watery fluid that bathes the villi with chyme particles. Fluid production results from local irritation of nerve cells and, possibly, from hormonal stimulation by secretin and cholecystokinin. The microvillous brush border secretes various hormones and digestive enzymes that catalyse final nutrient breakdown.

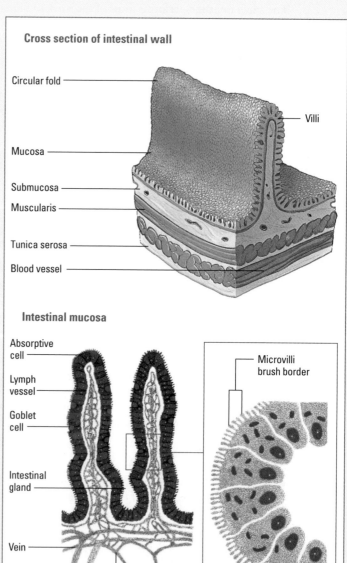

Cross section of intestinal wall

Circular fold
Villi
Mucosa
Submucosa
Muscularis
Tunica serosa
Blood vessel

Intestinal mucosa

Absorptive cell
Lymph vessel
Goblet cell
Intestinal gland
Vein
Artery
Microvilli brush border

Bacterial action

The large intestine harbours the bacteria *Escherichia coli*, *Enterobacter aerogenes*, *Clostridium perfringens* and *Lactobacillus bifidus*. All of these bacteria help synthesise vitamin K and break down cellulose into a usable carbohydrate. Bacterial action also produces *flatus*, which helps propel faeces towards the rectum.

Mucosa preparing for . . .

In addition, the mucosa of the large intestine produces *alkaline secretions* from tubular glands composed of goblet cells. This alkaline mucus lubricates the intestinal walls as undigested materials and wastes push through, protecting the mucosa from erosion and acidic bacterial action.

We're not all bad. Some of us bacteria aid GI function.

. . . mass movement

In the lower colon, long and relatively sluggish contractions cause propulsive waves, or *mass movements*. Normally occurring several times per day, these movements propel intestinal contents into the rectum and produce the urge to defecate.

Defecation normally results from the *defecation reflex*, a sensory and parasympathetic nerve-mediated response, along with the voluntary relaxation of the external anal sphincter. (See *GI changes with aging*.)

As time goes by . . .

GI changes with aging

The physiological changes that accompany aging usually prove less debilitating in the GI system than in most other body systems. Normal changes include diminished mucosal elasticity and reduced GI secretions, which, in turn, modify some processes—for example, digestion and absorption. GI tract motility, bowel wall and anal sphincter tone and abdominal muscle strength also may decrease with age. Any of these changes may cause complaints in an older patient, ranging from loss of appetite to constipation.

Changes in the oral cavity also occur. Tooth enamel wears away, leaving the teeth prone to cavities. Periodontal disease increases and the number of taste buds declines. The sense of smell diminishes and salivary gland secretion decreases, leading to appetite loss.

Liver changes

Normal physiological changes in the liver include decreased liver weight, reduced regenerative capacity and decreased blood flow to the liver. Because hepatic enzymes involved in oxidation and reduction markedly decline with age, the liver metabolises drugs and detoxifies substances less efficiently.

Quick quiz

1. Which component of the GI system completes food digestion?
   A. Stomach
   B. Gallbladder
   C. Small intestine
   D. Large intestine

Answer: C. The small intestine completes the process of digestion.

2. One of the functions of the liver is:
   A. regulating gastrin secretion.
   B. storing vitamins A, C and E.
   C. activation of vitamins K, D and B_{12}.
   D. detoxifying endogenous and exogenous toxins in plasma.

Answer: D. Among many other functions, the liver is responsible for detoxifying various exogenous and endogenous toxins in plasma.

3. Which GI hormone stimulates gastric secretion and motility?
   A. Gastrin
   B. Gastric inhibitory peptide
   C. Secretin
   D. Cholecystokinin

Answer: A. Gastrin is produced in the pyloric antrum and duodenal end mucosa and stimulates gastric secretion and motility.

4. In which phase of digestion does the stomach secrete the digestive juices hydrochloric acid and pepsin?
   A. Cephalic
   B. Gastric
   C. Intestinal
   D. Stomach emptying

Answer: A. By the time food is travelling towards the stomach, the cephalic phase—during which the stomach secretes digestive juices—has begun.

Scoring

★★★ If you answered all four questions correctly, excellent! You've passed through the GI system with the greatest of ease!

★★ If you answered three questions correctly, super! You've chewed the fat of this system and it's time to move on.

★ If you answered fewer than three questions correctly, don't worry! It might take a little longer to digest this material, but keep at it!

⓫ Nutrition and metabolism

Just the facts

In this chapter, you'll learn:

♦ roles of carbohydrates, proteins and lipids in nutrition

♦ functions of vitamins and minerals in the body

♦ the way in which glucose is turned into energy

♦ role of hormones in metabolism.

Nutrition

Nutrition refers to the intake, assimilation and utilisation of nutrients. The crucial nutrients in foods must be broken down into components for use by the body. Within cells, the products of digestion undergo further chemical reactions.

Metabolism refers to the sum of these chemical reactions. Through metabolism, food substances are transformed into energy or materials that the body can use or store.

Metabolism involves two processes:

 anabolism—synthesis of simple substances into complex ones

 catabolism—breakdown of complex substances into simpler ones releasing energy in the process.

Needed for nutrition

The body needs a continual supply of water and various nutrients for growth and repair. Virtually all nutrients come from digested food. The three major types of nutrients required by the body are carbohydrates, proteins and lipids.

Needed for metabolism

Vitamins are essential for normal metabolism. They contribute to the enzyme reactions that promote the metabolism of carbohydrates, proteins and lipids. *Minerals* are also important. They participate in such essential functions as enzyme metabolism and membrane transfer of essential elements.

Carbohydrates

Carbohydrates are organic compounds composed of carbon, hydrogen and oxygen (in the ratio 1:2:1) that are converted to glucose in the body; they yield approximately 4 kcal/g when used for energy.

> The energy in nutrients is measured in kilocalories—commonly just called calories. Adults need between 1,600 and 2,800 kcal daily, depending on their age, height, weight and physical activity.

Simple? Or complex?

Carbohydrates are categorised as simple or complex. Simple carbohydrates include the sugars in fruits, vegetables, dairy products and foods made with processed sugar. They raise the blood glucose level quickly. Complex carbohydrates include the starches and fibre found in breads, grains and beans. They raise the blood glucose level more slowly than simple carbohydrates.

Let's go for 'aride'

Sugars are classified as *monosaccharides*, *disaccharides* and *polysaccharides*. Sugars are carbohydrates and function as the body's primary energy source.

Monosaccharides
Monosaccharides are simple sugars that can't be split into smaller units by hydrolysis (a chemical reaction involving the removal of a molecule of water). The main monosaccharides are glucose, fructose and galactose.

Disaccharides
Disaccharides are synthesised from monosaccharides. A disaccharide molecule consists of two monosaccharides minus a water molecule. Examples of disaccharides include:
* *sucrose*, common table sugar, which is also found in some fruits and vegetables—a combination of a glucose molecule and a fructose molecule
* *lactose*, the sugar found in milk—a combination of a glucose molecule and a galactose molecule
* *maltose*, a sugar used in brewing and distilling—a combination of two glucose molecules

Polysaccharides

Like disaccharides, *polysaccharides* are synthesised from monosaccharides. A polysaccharide consists of a long chain (*polymer*) of more than 10 monosaccharides linked by glycoside bonds. Polysaccharides are ingested and broken down into simple sugars and then used as fuel molecules. There are two types of polysaccharide: starch from plants and glycogen, the storage form of glucose found in the liver. The body builds glycogen by using excess sugar (monosaccharides) and stores it for future use. When glycogen reserves are full, the liver converts the excess to fat.

Cellulose is another example of a polysaccharide, but it can't be easily broken down into simple sugars. Thus, the body derives little energy from this material, often referred to as dietary fibre. However, it is considered important in the prevention of conditions such as diverticulosis, constitpation, appendicitis, obesity and diabetes mellitus.

Proteins

Amino acids are the building blocks of proteins.

Proteins are complex nitrogenous organic compounds containing amino acid chains; some also contain sulphur and phosphorus. Proteins are used mainly for growth and repair of body tissues; when used for energy, they yield 4 kcal/g. Some proteins combine with lipids to form *lipoproteins* or with carbohydrates to form *glycoproteins*.

Amino acids

Amino acids are the building blocks of proteins. Each amino acid has a functional group to which a carboxyl (COOH) group and an amino (NH_2) group are attached.

The bonds that link . . .

Amino acids unite by through a dehydration reaction involving the COOH group on one amino acid with the amino group of the adjacent amino acid. This reaction releases a water molecule and creates a linkage called a *peptide bond*.

. . . and the acids that attract

The sequence and types of amino acids in the chain determine the nature of the protein. Each protein is synthesised on a ribosome as a straight chain. Chemical attractions between the amino acids in various parts of the chain cause the chain to coil or twist into a specific structure. A protein's structure largely determines its function.

Levels of structure

Structural features of proteins are usually described at four levels of complexity:

- *primary structure:* the linear arrangement of amino acids in a protein and the location of covalent linkages such as disulphide bonds between amino acids.
- *secondary structure:* areas of folding or coiling within a protein; examples include alpha helices and pleated sheets, which are stabilised by hydrogen bonding.
- *tertiary structure:* the final three-dimensional structure of a protein, which results from a large number of non-covalent interactions between amino acids.
- *quaternary structure:* non-covalent interactions that bind multiple polypeptides into a single, larger protein. Haemoglobin has quaternary structure due to association of two alpha globin and two beta globin polyproteins.

The tertiary and quaternary structures of proteins can be disrupted by the actions of pH, organic salts or heat.

Memory jogger

Remember the three **P**s of proteins:

Peptide = 2 to 10 amino acids
Polypeptide = 10 or more amino acids
Protein = 50 or more amino acids

Lipids

Lipids are organic compounds that don't dissolve in water but do dissolve in alcohol and other organic solvents. Lipids are a concentrated form of fuel and yield approximately 9 kcal/g when used for energy. The major lipids include fats (the most common type of lipid), phospholipids and steroids.

Fats

A fat, or *triglyceride,* contains three molecules of fatty acid combined with one molecule of glycerol. A fatty acid is composed of a chain of carbon atoms with hydrogen and a few oxygen atoms attached. Fatty acid chains vary in length. Long-chain fatty acids (12 or more carbon atoms) are found in most food fats.

The glycerol example

Glycerol, for example, is a three-carbon compound (alcohol) with an OH group attached to each carbon atom. The COOH group on each fatty acid molecule links to one OH group on the glycerol molecule; this results in the release of a water molecule.

Phospholipids

Phospholipids are complex lipids that are similar to fat but have a phosphorus- and nitrogen-containing compound that replaces one of the fatty acid molecules. Phospholipids are major structural components of cell membranes.

Steroids

Steroids are complex molecules in which the carbon atoms form four cyclic structures attached to various side chains. They contain no glycerol or fatty acid molecules. Examples of steroids include cholesterol, bile salts and sex hormones.

Vitamins and minerals

Vitamins are organic compounds that are needed in small quantities for normal metabolism, growth and development. Vitamins are classified as *water soluble* or *fat soluble*.

Daily routine

Water-soluble vitamins aren't stored in the body and must be replaced daily. Water-soluble vitamins include the B complex and C vitamins.

Eat up! The body can't manufacture enough of us on its own.

Stored but not forgotten

Fat-soluble vitamins are dissolved in fat before they are absorbed by the bloodstream. Excess fat-soluble vitamins are stored in the liver and body tissues; therefore, they don't need to be ingested daily. The fat-soluble vitamins include A, D, E and K. (See *Guide to vitamins and minerals*, page 194.)

Minerals

Minerals are inorganic substances that play important roles in:
- enzyme structure and function
- membrane transfer of essential compounds
- regulation of acid-base balance
- osmotic pressure
- muscle contractility
- nerve impulse transmission
- growth

Minerals are found in bones, haemoglobin, thyroxine, teeth and organs. They're classified as *major minerals* (more than 0.005% of body weight) or *trace minerals* (less than 0.005% of body weight). Major minerals include calcium, chloride, magnesium, phosphorus, potassium and sodium. Trace minerals include chromium, cobalt, copper, fluorine, iodine, iron, manganese, molybdenum, selenium and zinc.

(Text continues on page 196.)

Guide to vitamins and minerals

Good health requires intake of adequate amounts of vitamins and minerals to meet the body's metabolic needs. A vitamin or mineral excess or deficiency can lead to various disorders. This chart reviews major functions of vitamins and minerals and their food sources.

Vitamin or mineral	Major functions	Food sources
Water-soluble vitamins		
Vitamin C	• Collagen production, fine bone and tooth formation, iodine conservation, healing, red blood cell (RBC) formation, infection resistance	• Fresh fruits and vegetables
Vitamin B_1	• Carbohydrate metabolism, circulation, digestion, growth, learning ability, muscle tone maintenance, central nervous system (CNS) maintenance	• Meats, fish, poultry, pork, molasses, brewer's yeast, brown rice, nuts, wheat germ, whole and enriched grains
Vitamin B_2	• RBC formation; energy metabolism; cell respiration; epithelial, eye and mucosal tissue maintenance	• Meats, fish, poultry, milk, molasses, brewer's yeast, eggs, fruit, green leafy vegetables, nuts, whole grains
Vitamin B_6	• Antibody formation, digestion, deoxyribonucleic acid (DNA) and ribonucleic acid (RNA) synthesis, fat and protein utilisation, amino acid metabolism, haemoglobin production, CNS maintenance	• Meats, poultry, bananas, molasses, brewer's yeast, liver, fish, green leafy vegetables, peanuts, raisins, walnuts, wheat germ, whole grains
Folic acid	• Cell growth and reproduction, digestion, liver function, DNA and RNA formation, protein metabolism, RBC formation	• Citrus fruits, eggs, green leafy vegetables, milk products, offal, seafood, whole grains
Niacin	• Circulation, cholesterol level reduction, growth, hydrochloric acid production, metabolism (carbohydrate, protein, fat), sex hormone production	• Eggs, lean meats, milk products, offal, peanuts, poultry, seafood, whole grains
Vitamin B_{12}	• RBC formation, cellular and nutrient metabolism, tissue growth, nerve cell maintenance, appetite stimulation	• Beef, eggs, fish, milk products, offal, pork
Fat-soluble vitamins		
Vitamin A	• Body tissue repair and maintenance, infection resistance, bone growth, nervous system development, cell membrane metabolism and structure, night vision	• Meat, milk, eggs, butter • Leafy green and yellow vegetables, yellow fruits (sources of carotene—a precursor to vitamin A)
Vitamin D	• Calcium and phosphorus metabolism (bone formation), myocardial function, nervous system maintenance, normal blood clotting	• Egg yolks, offal, butter, cod liver oil, fatty fish
Vitamin E	• Aging retardation, anticlotting factor, diuresis, fertility, lung protection (antipollution), male potency, muscle and nerve cell membrane maintenance, myocardial perfusion, serum cholesterol reduction	• Butter, dark green vegetables, eggs, fruits, nuts, offal, vegetable oils, wheat germ

Guide to vitamins and minerals (continued)

Vitamin or mineral	Major functions	Food sources
Vitamin K	• Liver synthesis of prothrombin and other blood-clotting factors	• Green leafy vegetables, sunflower oil, yogurt, liver, molasses • Also manufactured by bacteria that line the GI tract

Minerals

Vitamin or mineral	Major functions	Food sources
Calcium	• Required for blood clotting, bone and tooth formation, cardiac rhythm, cell membrane permeability, muscle growth and contraction, nerve impulse transmission	• Cheese, milk, molasses, yogurt, whole grains, nuts, legumes, leafy vegetables
Chloride	• Maintenance of fluid, electrolyte, acid-base, and osmotic pressure balance	• Fruits, vegetables, table salt
Magnesium	• Acid-base balance, metabolism, protein synthesis, muscle relaxation, cellular respiration, nerve impulse transmission, important coenzyme activities	• Green leafy vegetables, nuts, seafood, cocoa, whole grains
Phosphorus	• Bone and tooth formation, cell growth and repair, energy production	• Eggs, fish, grains, meats, poultry, milk, milk products
Potassium	• Heartbeat, muscle contraction, nerve impulse transmission, rapid growth, fluid distribution and osmotic pressure balance, acid-base balance	• Seafood, molasses, vegetables, fruits, nuts
Sodium	• Cellular fluid level maintenance, muscle contraction, acid-base balance, cell permeability, muscle function, nerve impulse transmission	• Cheese, milk, salt, processed foods, canned soups, fast food, soy sauce
Fluoride (fluorine)	• Bone and tooth formation	• Fluoridated water, seafood
Iodine	• Thyroid hormone production, energy production, metabolism, physical and mental development	• Kelp, salt (iodised), seafood
Iron	• Important for growth in children, haemoglobin production, stress and disease resistance, cellular respiration, oxygen transport	• Egg yolks, meats, poultry, wheat germ, liver, oysters, enriched breads and cereals, green vegetables, molasses
Selenium	• Immune mechanisms, mitochondrial adenosine triphosphate synthesis, cellular protection	• Seafood, meats, liver, eggs
Zinc	• Burn and wound healing, carbohydrate digestion, metabolism (carbohydrate, fat, protein), prostate gland function, reproductive organ growth and development, cell growth	• Liver, mushrooms, seafood, soybeans, spinach, meats

Digestion and absorption

Nutrients must be digested in the GI tract by enzymes that split large units into smaller ones. In this process, called *hydrolysis*, a compound unites with water and then splits into simpler compounds. The smaller units are then absorbed from the small intestine and transported to the liver through the portal venous system.

> Time to split! During hydrolysis, a compound unites with water and then splits into simpler compounds.

Carbohydrate digestion and absorption

Enzymes break down complex carbohydrates. In the oral cavity, *salivary amylase* initiates starch hydrolysis into disaccharides. In the small intestine, *pancreatic amylase* continues this process.

Splitting, hydrolysing . . .

Disaccharide enzymes in the intestinal mucosa hydrolyse disaccharides into monosaccharides. Lactase splits the compound lactose into glucose and galactose, and sucrase hydrolyses the compound sucrose into glucose and fructose.

. . . and movin' along

Monosaccharides, such as glucose, fructose and galactose, are absorbed through the intestinal mucosa and are then transported through the portal venous system to the liver. There, enzymes convert fructose and galactose to glucose.

Ribonucleases and deoxyribonucleases break down nucleotides from deoxyribonucleic acid and ribonucleic acid into pentoses and nitrogen-containing compounds (nitrogen bases). Like glucose, these compounds are absorbed through the intestinal mucosa.

Protein digestion and absorption

Enzymes digest proteins by hydrolysing the peptide bonds that link the amino acids of the protein chains. This process of hydrolysis restores water molecules.

Gastric pepsin breaks proteins into polypeptides. Pancreatic trypsin, chymotrypsin and carboxypeptidases convert polypeptides into peptides

The breakdown lane

Intestinal mucosal peptidases break down peptides into their constituent amino acids. After being absorbed through the intestinal mucosa by active transport mechanisms, these amino acids travel through the portal venous system to the liver. The liver converts the amino acids not needed for protein synthesis into glucose.

Lipid digestion and absorption

Fat digestion occurs in the small intestine. Pancreatic lipase breaks down fats and phospholipids into a mixture of glycerol, short- and long-chain fatty acids and monoglycerides. The portal venous system then carries these substances to the liver. Lipase hydrolyses the bonds between glycerol and fatty acids—a process that restores the water molecules released when the bonds were formed.

> Amino acids:
> Enter here and
> prepare to
> convert.

On a short leash . . .

Glycerol diffuses directly through the mucosa. Short-chain fatty acids diffuse into the intestinal epithelial cells and are carried to the liver via the portal venous system.

. . . or on a long one

Long-chain fatty acids and monoglycerides in the intestine dissolve in the bile salt micelles and then diffuse into the intestinal epithelial cells. There, lipase breaks down absorbed monoglycerides into glycerol and fatty acids. In the smooth endoplasmic reticulum of the epithelial cells, fatty acids and glycerol recombine to form fats.

Chylomicrons

Along with a small amount of cholesterol and phospholipid, triglycerides are coated with a thin layer of protein to form lipoprotein particles called *chylomicrons*. Chylomicrons collect in the intestinal lacteals (lymphatic vessels) and are carried through lymphatic channels. After entering the circulation through the thoracic duct, they're distributed to body cells.

Stored away for later

In the cells, fats are extracted from the chylomicrons and broken down by enzymes into fatty acids and glycerol. Then they're absorbed and recombined in fat cells, reforming triglycerides for storage and later use.

Carbohydrate metabolism

Carbohydrates are the preferred energy fuel of human cells. Most of the carbohydrates in absorbed food is quickly catabolised for the release of energy.

Glucose to energy

All ingested carbohydrates are converted to glucose, the body's main energy source. Glucose not needed for immediate energy is stored as glycogen or converted to lipids.

Energy from glucose catabolism is generated in three phases:

This whole nutrition and metabolism thing is pretty simple; it's about getting energy to do things!

- glycolysis

- Krebs cycle (also called the *citric acid cycle*)

- the electron transport system. (See *Tracking the glucose pathway*.)

Glycolysis, which occurs in the cell cytoplasm, doesn't use oxygen. The other two phases, which occur in mitochondria, do use oxygen.

Glycolysis

Glycolysis refers to the process by which enzymes break down the 6-carbon glucose molecule into two 3-carbon molecules of pyruvic acid (pyruvate). Glycolysis produces a small amount of energy in the form of adenosine triphosphate (ATP).

Cruising the glucose pathway

Next, pyruvic acid releases a carbon dioxide (CO_2) molecule and is converted in the mitochondria to a two-carbon acetyl fragment, which combines with a complex organic compound called *coenzyme A* (CoA) to form acetyl CoA.

Krebs cycle

The second phase in glucose catabolism is the Krebs cycle. It's the pathway by which a molecule of acetyl CoA is oxidised by enzymes to yield energy. It similarly produces a small amount of ATP.

Carbons, carbons everywhere

The two-carbon acetyl fragments of acetyl CoA enter the Krebs cycle by joining to the four-carbon compound oxaloacetic acid to form citric acid, a six-carbon compound. In this process, the CoA molecule detaches from the acetyl group, becoming available to form more acetyl CoA molecules. Enzymes convert citric acid into intermediate compounds and eventually convert it back into oxaloacetic acid. Then the cycle can begin again.

In addition to liberating CO_2 and generating energy, each turn of the Krebs cycle releases hydrogen atoms, which are picked up by the coenzymes nicotinamide adenine dinucleotide (NAD) and flavin adenine dinucleotide (FAD).

Now I get it!

Tracking the glucose pathway

Glucose catabolism generates energy in three phases: glycolysis, Krebs cycle and the electron transport system. This flowchart summarises the first two phases.

Glycolysis

Glycolysis, the first phase, breaks apart one molecule of glucose to form two molecules of pyruvate, which yields energy in the form of adenosine triphosphate and acetyl coenzyme A (CoA).

Krebs cycle

The second phase, the Krebs cycle, continues carbohydrate metabolism. Fragments of acetyl CoA join to oxaloacetic acid to form citric acid. The CoA molecule breaks off from the acetyl group and may form more acetyl CoA molecules. Citric acid is first converted into intermediate compounds and then back into oxaloacetic acid. The Krebs cycle also liberates carbon dioxide.

Electron transport system

In the third phase of glucose catabolism, molecules on the inner mitochondrial membrane attract electrons from hydrogen atoms and carry them through oxidation-reduction reactions in the mitochondria. The hydrogen ions produced in the Krebs cycle then combine with oxygen to form water.

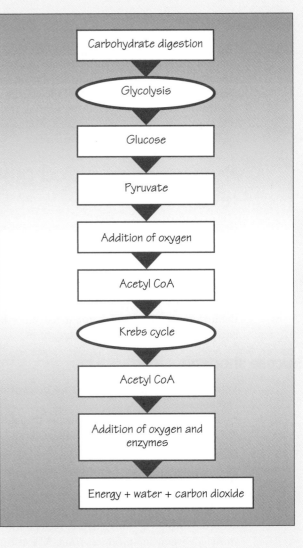

Glucose is the body's main energy source.

Electron transport system

The electron transport system is the last phase of carbohydrate catabolism. In this phase, carrier molecules on the inner mitochondrial membrane pick up electrons from the hydrogen atoms carried by NAD and FAD. (Each hydrogen atom contains a hydrogen ion and an electron.) These carrier molecules transport the electrons through a series of enzyme-catalysed oxidation-reduction reactions in the mitochondria. The electron transport chain produces large amounts of ATP.

> The last step of carbohydrate catabolism is electron transport.

Oxygen attraction

Oxygen plays a crucial role by attracting electrons along the chain of carriers in the transport system. During *oxidation*, a chemical compound loses electrons; during *reduction*, it gains electrons. These reactions release the energy contained in the electrons and generate ATP. After passing through the electron transport system, the hydrogen ions produced in the Krebs cycle combine with oxygen to form water.

Regulation of blood glucose levels

Because all ingested carbohydrates are converted to glucose, the body depends on the liver, muscle cells and certain hormones to regulate blood glucose levels.

Liver

When glucose levels exceed the body's immediate needs, hormones stimulate the liver to convert glucose into glycogen or lipids. Glycogen forms through *glycogenesis*; lipids form through *lipogenesis*.

Glucose shortage

When the blood glucose level drops excessively, the liver can form glucose by two processes:
- breakdown of glycogen to glucose through glycogenolysis
- synthesis of glucose from amino acids and glycerol through gluconeogenesis.

> I convert glucose into glycogen or lipids.

Muscle cells

Muscle cells can convert glucose to glycogen for storage. However, they lack the enzymes to convert glycogen back to glucose when needed. During vigorous exercise skeletal muscle tissue breaks down stored glycogen and produces some ATP through anaerobic glycolysis. The pyruvate that results is converted to yield lactic acid and a small amount of energy. Lactic acid then builds up in the muscles, and muscle glycogen is eventually depleted.

Now I get it!

How insulin affects blood glucose level

Unlike most hormones, insulin tends to decrease the blood glucose level. It does this by helping glucose to enter cells, thereby promoting glycogenesis and stimulating glucose catabolism (glycolysis).

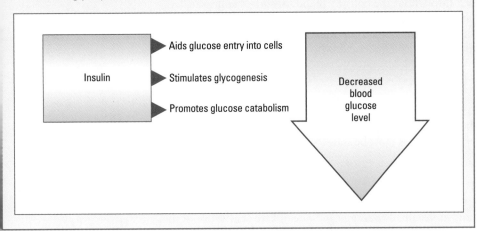

Insulin

Aids glucose entry into cells

Stimulates glycogenesis

Promotes glucose catabolism

Decreased blood glucose level

Look out glycogen! I'm breaking you down and creating energy!

Glycogen returns

Some of the lactic acid diffuses from muscle cells, is transported to the liver and is converted to pyruvate or glucose which travels through the bloodstream to the muscles and reforms into glycogen.

Energise!

When muscle exertion stops, some of the accumulated lactic acid converts back to pyruvic acid. Pyruvic acid is oxidised completely to yield energy by means of the Krebs cycle and the electron transport system.

Hormones

Hormones regulate the blood glucose level by stimulating the metabolic processes that restore a normal level in response to blood glucose changes. (See *How insulin affects blood glucose level*.)

Sugar shift

Insulin, produced by the pancreatic islet cells, is the only hormone that significantly reduces the blood glucose level. In addition to promoting cell uptake and use of glucose as an energy source, insulin promotes glucose storage as glycogen (glycogenesis) and lipids (lipogenesis). Therefore, insulin production has widespread effects throughout the body.

Protein metabolism

Proteins are absorbed as amino acids and carried by the portal venous system to the liver, and then throughout the body by blood. Absorbed amino acids mix with other amino acids in the body's amino acid pool. These other amino acids may be synthesised by the body from other substances, such as *keto acids*, or they may be produced by protein breakdown.

Amino acid conversion

The body can't store amino acids; instead, it converts them to protein or glucose or catabolises them to provide energy. Before these changes can occur, however, amino acids must be transformed by deamination or transamination.

Deamination
In *deamination*, an amino group ($-NH_2$) splits off from an amino acid molecule to form one molecule of ammonia and one of keto acid. Most of the ammonia is converted to urea and excreted in urine.

Transamination
In *transamination*, an amino group is exchanged for a keto group in a keto acid through the action of transaminase enzymes. During this process, the amino acid is converted to a keto acid and the original keto acid is converted to an amino acid.

Amino acid synthesis

Proteins are synthesised from 20 amino acids from the body's amino acid pool. (See *Essential and nonessential amino acids*.)

Power station

Amino acids not used for protein synthesis can be converted to keto acids and metabolised by the Krebs cycle and the electron transport system to produce energy.

Fat chance

Amino acids not used for protein synthesis may be converted to pyruvic acid and then to acetyl CoA. The acetyl CoA fragments condense to form

Essential and nonessential amino acids

Amino acids are the structural units of proteins. They're classified as essential or nonessential based on whether the human body can synthesise them. The nine essential amino acids that can't be synthesised must be obtained from the diet. The other 11 can be synthesised and are therefore nonessential in the diet; however, they're needed for protein synthesis.

Essential
- Histidine
- Isoleucine
- Leucine
- Lysine
- Methionine
- Phenylalanine
- Threonine
- Tryptophan
- Valine

Nonessential
- Alanine
- Arginine
- Asparagine
- Aspartic acid
- Cystine
- Glutamine
- Glycine
- Hydroxyproline
- Proline
- Serine
- Tyrosine

There are 20 amino acids: 9 essential and 11 nonessential.

long-chain fatty acids—a process that's the reverse of fatty acid breakdown. These fatty acids then combine with glycerol to form fats.

Going glucose

Amino acids can also be converted to glucose. They're first converted to pyruvic acid, which may then be converted to glucose.

Lipid metabolism

Lipids are stored in adipose tissue within cells until they're required for use as fuel. When needed for energy, each fat molecule is hydrolysed to glycerol and three molecules of fatty acids. Glycerol can be converted to pyruvic acid and then to acetyl CoA, which enters the Krebs cycle.

Ketone body formation

The liver normally forms ketone bodies from acetyl CoA fragments, derived largely from fatty acid catabolism. Acetyl CoA molecules yield three types of ketone bodies: acetoacetic acid, beta-hydroxybutyric acid and acetone.

As time goes by . . .

Nutrition-related changes with aging

As a person ages, caloric needs decrease. Protein, vitamin and mineral requirements usually remain the same throughout life. The body's ability to process these nutrients, however, is also affected by the aging process.

Physiological changes

Diminished intestinal motility typically accompanies aging and may cause constipation. Physical inactivity, emotional stress, medications and nutritionally inadequate diets of soft, refined foods that are low in dietary fibre can also cause constipation. Laxative abuse results in the rapid transport of food through the GI tract, decreasing digestion and absorption. Faecal incontinence may also occur in elderly patients.

   Other physiological changes that can affect nutrition in an older patient include:
- decreased renal function, causing greater susceptibility to dehydration and formation of renal calculi
- loss of calcium and nitrogen (in people who aren't ambulatory)
- diminished enzyme activity and gastric secretions

- reduced pepsin and hydrochloric acid secretion, which tends to diminish the absorption of calcium and vitamins B_1 and B_2
- decreased salivary flow and diminished sense of taste, which may reduce the appetite and increase a person's consumption of salty, sweet and spicy foods
- diminished intestinal motility and peristalsis of the large intestine
- thinning of tooth enamel, causing teeth to become more brittle
- decreased biting force
- diminished gag reflex.

Affecting factors

Nutritional status can be affected by such socioeconomic and psychological factors as loneliness, decline of the older person's importance in the family, susceptibility to nutritional fads and lack of money or transportation, which limit access to nutritious foods. In addition, some conditions that are common among older people can affect mobility and, therefore, the ability to obtain or prepare food or feed oneself.

Acetoacetic acid results from the combination of two acetyl CoA molecules and the subsequent release of CoA from these molecules.

   Beta-hydroxybutyric acid forms when hydrogen is added to the oxygen atom in the acetoacetic acid molecule. The term *beta* indicates the location of the carbon atom containing the OH group.

   Acetone forms when the COOH group of acetoacetic acid releases CO_2. Muscle tissue, brain tissue and other tissues oxidise these ketone bodies for energy.

Excessive ketone formation

Under certain conditions, the body produces more ketone bodies than it can oxidise for energy. Such conditions include fasting, starvation and uncontrolled diabetes (in which the body can't break down glucose). The body must then use fat, rather than glucose, as its primary energy source.

Along with other tissues, I oxidise ketone bodies for energy.

Ketone cops!

Use of fat instead of glucose for energy leads to an excess of ketone bodies. This condition disturbs the body's normal acid-base balance and homeostatic mechanisms, leading to ketosis.

Lipid formation

Excess amino acids can be converted to fat through keto acid--acetyl CoA conversion. Glucose may be converted to pyruvic acid and then to acetyl CoA, which is converted into fatty acids and then fat (in much the same way that amino acids are converted into fat). (See *Nutrition-related changes with aging.*)

Quick quiz

1. Which type of nutrient yields 9 kcal/g when used for energy?
    A. Proteins
    B. Carbohydrates
    C. Lipids
    D. Vitamins

Answer: C. Lipids are a concentrated form of fuel and yield 9 kcal/g.

2. Which hormone decreases the blood glucose level?
    A. Epinephrine
    B. Cortisol
    C. Insulin
    D. Glycogen

Answer: C. Insulin is the only hormone that significantly reduces blood glucose. It does so by aiding glucose entry into cells, thereby stimulating glycogenesis and promoting glucose catabolism.

3. Essential amino acids are:
    A. organic compounds that are needed in small amounts for normal metabolism, growth and development.
    B. organic compounds that don't dissolve in water but do dissolve in alcohol and other organic solvents.
    C. the structural unit of protein that doesn't need to be obtained from the diet.
    D. the structural unit of protein that must be obtained from the diet.

Answer: D. Essential amino acids can't be synthesised in the body and, therefore, must be obtained from the diet.

4.   Which vitamin is involved in prothrombin synthesis and other blood-clotting factors?
    A.   Vitamin K
    B.   Vitamin E
    C.   Vitamin B_{12}
    D.   Vitamin B_6

Answer: A. Vitamin K is involved in liver synthesis of prothrombin and other blood-clotting factors.

Scoring

☆☆☆   If you answered all four questions correctly, hooray! You've digested a healthy dose of dense clinical material.

☆☆   If you answered three questions correctly, remarkable! You'll soon be a master of metabolism.

☆   If you answered fewer than three questions correctly, get energised! Just a few more quick quizzes to go!

12 Urinary system

Just the facts

In this chapter, you'll learn:

♦ major structures of the urinary system

♦ functions of the kidneys, ureters, bladder, and urethra

♦ the way in which urine is formed

♦ role of hormones in the urinary system.

Structures of the urinary system

The urinary system consists of:
- two kidneys
- two ureters
- the bladder
- the urethra.

Working together, these structures remove wastes from the body, help to govern acid-base balance by retaining and excreting hydrogen ions, regulate fluid and electrolyte balance and assist in blood pressure control.

> Think of us as the body's plumbers. Together with the ureters, bladder and urethra, we do everything from removing wastes . . .

> . . . to balancing fluids and electrolytes. We also keep blood pressure in check!

Kidneys

The *kidneys* are highly vascular organs located on either side of the vertebral column, between vertebrae T_{12} and L_3. They are retroperitoneal organs that each consist of three regions:

- renal cortex (outer region)
- renal medulla (middle region)
- renal pelvis (inner region). (See *A close look at the urinary system*.)

Three regions

- *The renal cortex (outer region) contains about 1.25 million renal tubules.*
- *The renal medulla (middle region) functions as a collecting chamber.*
- *The renal pelvis (inner region) receives urine through the major calyces.*

Filter station

The *renal cortex*, the outer region, contains blood-filtering structures and is protected by a fibrous capsule and layers of fat.

Renal wonder

The *renal medulla*, the middle region of the kidney, contains 8 to 12 renal pyramids—striated wedges that are composed mostly of tubular structures. The tapered portion of each pyramid empties into a cup-like *calyx*. These calyces channel formed urine from the pyramids into the *renal pelvis*.

Keeping kidneys safe

The kidneys are protected in front by the contents of the abdomen and behind by the muscles attached to the vertebral column. A layer of fat surrounding each kidney offers further protection.

Adrenal influence

On top of each kidney lies an adrenal gland. These glands are affected by the release of renin from the kidneys, which in turn affects the renal system by influencing blood volume and pressure through its positive effect on sodium and water retention.

No shortage of blood

Each kidney is supplied with blood by a renal artery, which subdivides into several branches when it enters the kidney. The kidneys are highly vascular structures receiving about 20% of the blood pumped by the heart each minute.

All in a day's work

Together, these tissues allow the kidneys to perform their many functions, including:

- elimination of waste and excess ions (in the form of urine)
- blood filtration (by regulating chemical composition and blood volume)

(Text continues on page 210.)

Zoom in

A close look at the urinary system

The kidneys are retroperitoneal organs located in the lumbar area, with the right kidney situated slightly lower than the left to make room for the liver (which lies just above it). The position of the kidneys shifts somewhat with changes in body position. Covering the kidneys are the true or fibrous capsule, perirenal fat and renal fasciae.

Blood's cleansing journey

The kidneys receive waste-filled blood from the renal artery, which branches off the abdominal aorta. After passing through a complicated network of smaller blood vessels and nephrons, the filtered blood returns to the circulation by way of the renal veins, which empty into the inferior vena cava.

Continuing the cleanup

The kidneys excrete waste products that the nephrons remove from the blood; these excretions combine with other waste fluids (such as urea, creatinine, phosphates and sulphates) to form urine. An action called *peristalsis* (the circular contraction and relaxation of a tube-shaped structure) passes the urine through the ureters and into the urinary bladder. When the bladder has filled, nerves in the bladder wall relax the sphincter. In conjunction with a voluntary stimulus, this relaxation causes urine to pass into the urethra for elimination from the body.

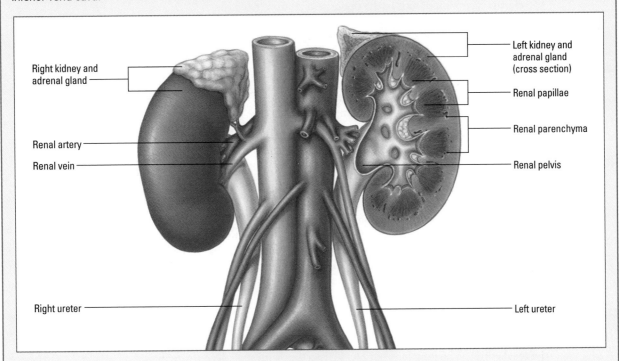

Right kidney and adrenal gland

Renal artery

Renal vein

Right ureter

Left kidney and adrenal gland (cross section)

Renal papillae

Renal parenchyma

Renal pelvis

Left ureter

- maintenance of fluid-electrolyte and acid-base balances
- production and release of renin to promote angiotensin II activation and aldosterone production in the adrenal gland
- production of erythropoietin (a hormone that stimulates red blood cell [RBC] production) and enzymes (such as renin, which helps govern blood pressure and kidney function)
- perform gluconeogenesis
- conversion of vitamin D to a more active form

The nephron

Within the kidney, the *nephron* serves as the basic structural and functional unit.

Filter, reabsorb and secrete

The nephrons perform three main functions:
- mechanically filtrating fluids, wastes, electrolytes, acids and bases into the tubular system
- reabsorption of molecules from this ultrafiltrate into the blood in response to specific body needs
- selective secretion of ions in the opposite direction (from the blood into the urine), allowing precise control of fluid and electrolyte balance

The nephrons mechanically filter fluids, wastes, electrolytes, acids and bases into the tubular system.

RENAL WASTE

Part and parcel

Each nephron consists of a glomerulus, a network of capillaries contained within a capsular structure (glomerular capsule), and a tubule.

Totally tubular

The nephron is divided into three portions. The portion nearest the glomerular capsule is the *proximal convoluted tubule*. The second portion, the *loop of Henlé*, has a descending and an ascending limb. The third portion, the one farthest from the glomerular capsule, is the *distal convoluted tubule*. Its distal end joins the far ends of neighbouring nephrons, forming a larger collecting tubule. (See *Structure of the nephron*.)

Positively loopy

The glomeruli and proximal and distal tubules of the nephron are located in the renal cortex. The long loops of Henlé, together with their accompanying blood vessels and collecting tubules, form the renal pyramids in the medulla.

Zoom in

Structure of the nephron

The *nephron* is the kidney's basic functional unit and the site of urine formation. The renal artery, a large branch of the abdominal aorta, carries blood to each kidney. Blood flows through the interlobular artery (running between the lobes of the kidneys) to the afferent arteriole, which conveys blood to the glomerulus. Blood passes through the glomerulus into the efferent arteriole and into the peritubular capillaries, venules and the interlobular vein. The peritubular capillary network of vessels then supplies blood to the tubules of the nephron. Blood is additionally supplied to the juxtaglomerular capillaries by a network of vessels termed the *vasa recta*. This type of nephron has an added role in the concentration of urine.

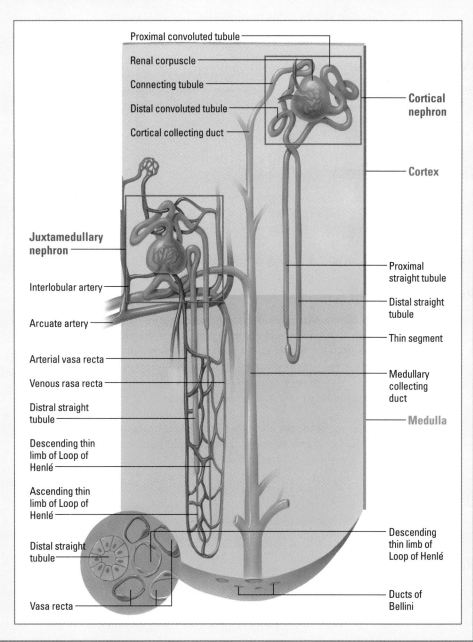

Proximal convoluted tubule

Renal corpuscle

Connecting tubule

Distal convoluted tubule

Cortical collecting duct

Cortical nephron

Cortex

Juxtamedullary nephron

Interlobular artery

Arcuate artery

Arterial vasa recta

Venous rasa recta

Distral straight tubule

Descending thin limb of Loop of Henlé

Ascending thin limb of Loop of Henlé

Distal straight tubule

Vasa recta

Proximal straight tubule

Distal straight tubule

Thin segment

Medullary collecting duct

Medulla

Descending thin limb of Loop of Henlé

Ducts of Bellini

About 80% to 85% of the nephrons have short loops of Henlé that only penetrate into the superficial region of the renal medulla. These nephrons usually have their glomeruli in the superficial region of the renal cortex and are called *cortical nephrons*. The remaining nephrons (15% to 20%) are called juxtamedullary nephrons. These structures have glomeruli deep in the renal cortex close to the renal medulla, and long loops of Henlé that transverse the renal medulla almost to the renal papilla. The juxtaglomerular nephrons have a role in the production of very dilute or concentrated urine.

Water, water everywhere

The proximal convoluted tubules have freely permeable cell membranes. This allows reabsorption of nearly all the filtrate's glucose, amino acids, metabolites and electrolytes into nearby capillaries as well as allowing for the reabsorption of large amounts of water.

Time to concentrate

By the time the filtrate enters the descending limb of the loop of Henlé, its water content has been reduced by 70%. At this point, the filtrate contains a high concentration of salts, chiefly sodium. As the filtrate moves deeper into the medulla and the loop of Henlé, osmosis draws even more water into the extracellular spaces, further concentrating the filtrate.

Readjust and exit

After the filtrate enters the ascending limb, its concentration is readjusted by the transport of sodium and chloride ions out of the tubule. This transport continues until the filtrate enters the distal convoluted tubule.

Ureters

The *ureters* are fibromuscular tubes that connect each kidney to the bladder. Because the left kidney is higher than the right kidney, the left ureter is usually slightly longer than the right ureter.

Triple protection

Each ureter is surrounded by a three-layered wall. (See *Three layers of the ureter*.)

Riding the waves

The ureters act as conduits that carry urine from the kidneys to the bladder. Peristaltic waves occurring one to five times each minute channel urine along the ureters toward the bladder.

Three layers of the ureter

Each ureter has a three-layered wall:
- The *mucosa*, the innermost layer, contains the transitional epithelium.
- The *muscularis*, the middle layer, contains smooth muscle layers.
- Extensions of the *fibrous coat*, the outer layer, hold the ureter in place.

Bladder

The *bladder* is a hollow, muscular organ in the pelvis. It is located in the pelvic cavity, posterior to the *symphysis pubis* (the joint between the two pubic bones). Its function is to store urine. In a normal adult, bladder capacity ranges from 500 to 600 ml although most individuals would feel the need to void when the volume reaches approximately 300 ml. If the amount of stored urine exceeds bladder capacity, the bladder distends above the symphysis pubis.

Urination is the result of two processes—one voluntary and one involuntary.

Good things come in threes

The base of the bladder contains three openings that form a triangular area called the *trigone*. Two of the openings connect the bladder to the ureters, while the third connects the bladder to the urethra.

A matter of reflex

Urination results from involuntary (reflex) and voluntary (learned or intentional) processes. When urine fills the bladder, parasympathetic nerve fibres in the bladder wall cause the bladder to contract and the *internal sphincter* (located at the internal urethral orifice) to relax. The cerebrum, in a voluntary reaction, then causes the external sphincter to relax and urination to begin. This is called the *micturition* reflex.

Urethra

The *urethra* is a muscular tube that channels urine from the bladder to the outside of the body.

Female connections

In the female, the urethra is embedded in the anterior wall of the vagina behind the symphysis pubis. The urethra connects the bladder with an external opening, or *urethral meatus*, located anterior to the vaginal opening.

The female urethra is composed of an inner layer of mucous membrane, a middle layer of spongy tissue and an outer layer of muscle.

Male extensions

In the male, the urethra passes vertically through the *prostate* gland and then extends through the urogenital diaphragm and penis. The male urethra serves as a passageway for semen as well as urine.

Urine formation

Urine formation is one of the main functions of the urinary system. Urine formation results from three processes that occur in the nephrons: glomerular filtration, tubular reabsorption and tubular secretion. (See *How the kidneys form urine*.)

A mine of minerals

When formed, normal urine consists of water, sodium, chloride, potassium, calcium, magnesium, sulphates, phosphates, bicarbonates, uric acid, ammonium ions, creatinine and *urobilinogen* (a derivative of bilirubin resulting from the action of intestinal bacteria). Some drugs and their metabolites may also be excreted in the urine. Leucocytes and RBCs may be found in urine of patients with renal disease.

Kidneys in charge

The kidneys can vary the amount of substances reabsorbed and secreted in the nephrons, changing the composition of excreted urine.

Controlling the flow

Total daily urine output averages 800 to 2,100 ml, varying with fluid intake and climate. For example, after drinking a large volume of fluid, urine output increases as the body rapidly excretes excess water. If an individual restricts or decreases water intake or ingests excessive amounts of sodium, urine output decreases as the body retains water to restore normal fluid concentration.

We'd better get busy! We process up to 180 L of water every day!

Hormones and the urinary system

Hormones play a major role in the urinary system, including helping the body to manage tubular reabsorption and secretion. Hormones associated with the urinary system include:
- antidiuretic hormone
- angiotensin I
- angiotensin II
- aldosterone
- erythropoietin

Now I get it!

How the kidneys form urine

Urine formation occurs in three steps: glomerular filtration, tubular reabsorption and tubular secretion.

Step 1: Filter

As blood flows into the glomerulus, filtration occurs. Filtration is the movement of water (180 L per day) and all material present in blood (with the exception of the formed elements and most plasma proteins) from the glomerular capillaries into the glomerular capsule, under the pressure of blood.

Step 2: Reabsorb

Reabsorption occurs along the length of the tubule although most occurs in the proximal convoluted tubule.

Water (approximately 70% of the filtrate volume), sodium (Na^+), potassium (K^+), glucose and amino acids are reabsorbed in this region; Na^+, K^+, Cl^- and water are reabsorbed into the blood from the loop of Henlé; Na^+ and water are reabsorbed from the distal convoluted tubule under the influence of aldosterone; and water from the collecting ducts under the influence of antidiuretic hormone.

Step 3: Secrete

In tubular secretion, a substance moves from the peritubular capillaries into the tubular filtrate. Peritubular capillaries then secrete K^+, ammonia (much of it as NH_4^+) and other waste products into the distal tubules via active transport.

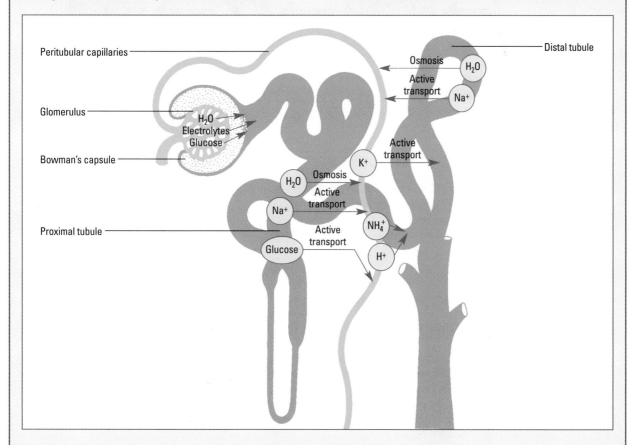

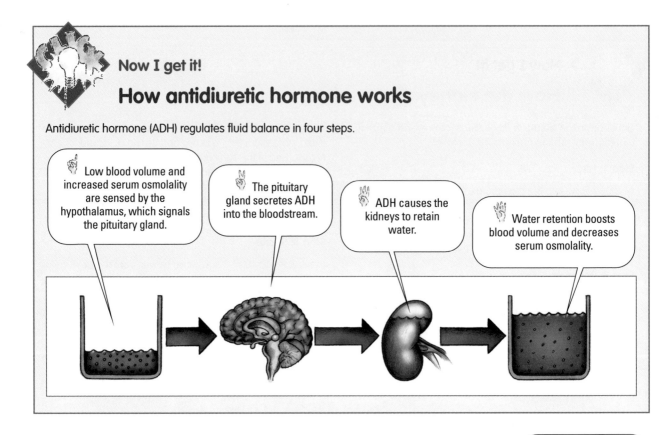

Now I get it!

How antidiuretic hormone works

Antidiuretic hormone (ADH) regulates fluid balance in four steps.

Low blood volume and increased serum osmolality are sensed by the hypothalamus, which signals the pituitary gland.

The pituitary gland secretes ADH into the bloodstream.

ADH causes the kidneys to retain water.

Water retention boosts blood volume and decreases serum osmolality.

Antidiuretic hormone

Antidiuretic hormone (ADH) regulates levels of urine output. High levels of ADH increase water reabsorption and urine concentration, whereas lower levels of ADH decrease water absorption and dilute urine. (See *How antidiuretic hormone works*.)

Renin-angiotensin system

Renin is an enzyme that's secreted by the kidneys and circulated in the blood. Renin itself has no effect on blood pressure, but it leads to the formation of the hormone called *angiotensin I*. As it circulates through the lungs, angiotensin I is converted into *angiotensin II* by *angiotensin-converting enzyme*. Angiotensin II exerts a powerful constricting effect on the arterioles. In this way, it can raise blood pressure.

High levels of ADH increase water reabsorption and urine concentration, whereas lower levels of ADH decrease water absorption and dilute urine.

The best defence

The primary function of the renin-angiotensin system is to serve as a defence mechanism, maintaining blood pressure in situations such as haemorrhage and extreme sodium depletion. Low blood pressure and low levels of sodium passing through the kidneys are two of the three factors that stimulate the kidneys to release renin. (The third is stimulation of the sympathetic nervous system.)

Aldosterone

The renin-angiotensin system has a second effect that makes it even more potent; it acts on the adrenal gland to release *aldosterone*.

BP assist

Aldosterone, produced by the adrenal cortex, causes the reabsorption of sodium ions (Na^+) from the distal convoluted tubule into the circulation; water follows by osmosis, which raises blood pressure. Aldosterone is also responsible for regulating potassium ion (K^+) levels indirectly since a K^+ is excreted when a Na^+ is absorbed. (See *The rennin-angiotensin-aldosterone system*, page 218.)

Age-related changes to blood pressure and flow significantly affect renal function. (See *Urinary changes with aging*, page 219.)

Erythropoietin

The kidneys secrete the hormone *erythropoietin* in response to low arterial oxygen tension. This hormone travels to the bone marrow, where it stimulates increased RBC production.

A balancing act

The kidneys also regulate calcium and phosphorus balance by filtering and reabsorbing approximately half of unbound serum calcium. In addition, the kidneys activate vitamin D_3, a compound that promotes intestinal calcium absorption and regulates phosphate excretion.

Now I get it!

The renin-angiotensin-aldosterone system

The renin-angiotensin-aldosterone system regulates the body's sodium and water levels and blood pressure. Juxtaglomerular cells (1) near the glomeruli in each kidney secrete the enzyme renin into the blood.

Renin circulates throughout the body and converts angiotensinogen, made in the liver (2), into angiotensin I. In the lungs (3), angiotensin I is converted by hydrolysis to angiotensin II. Angiotensin II acts on the adrenal cortex (4) to stimulate production of the hormone aldosterone. Aldosterone acts on the juxtaglomerular cells to increase sodium and water retention and to stimulate or depress further renin secretion, completing the feedback system that automatically readjusts homeostasis.

Angiotensin II also stimulates the release of antidiuretic hormone from the posterior pituitary leading to the reabsorption of water, vasoconstriction of blood vessels and increasing thirst.

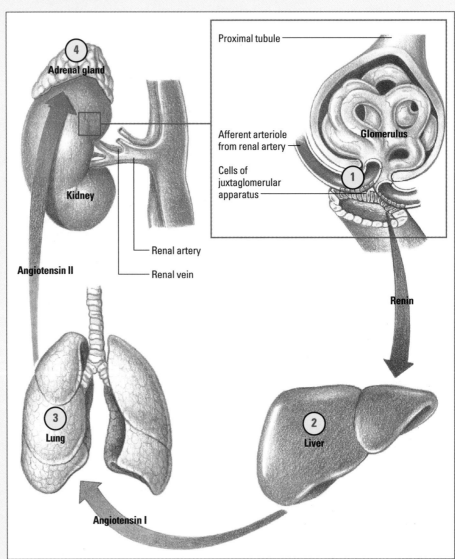

As time goes by . . .

Urinary changes with aging

As a person ages, changes in the kidneys and bladder can affect urinary system function.

Kidneys

After age 40, kidney function may diminish; by age 90, it may decrease by as much as 50%. Age-related changes in kidney vasculature that disturb glomerular haemodynamics result in a decline in glomerular filtration rate. Reduced cardiac output and age-related atherosclerotic changes cause kidney blood flow to decrease by up to 55%. In addition, tubular reabsorption and renal concentrating ability decline because the size and number of functioning nephrons decrease. Also, as blood levels of aldosterone and renin fall, the kidneys are less responsive to antidiuretic hormone.

Bladder

As a person ages, bladder muscles weaken. This may lead to incomplete bladder emptying and chronic urine retention—predisposing the bladder to infection.

And the rest

Other age-related changes that affect renal function include diminished kidney size, impaired renal clearance of drugs, reduced bladder size and capacity and decreased renal ability to respond to variations in sodium intake. By age 70, blood urea levels rise by 20%. Residual urine, frequency of urination and nocturia also increase with age.

Quick quiz

1. In a normal adult, bladder capacity ranges from:
    A. 50 to 100 ml.
    B. 200 to 300 ml.
    C. 500 to 600 ml.
    D. 700 to 900 ml.

Answer: C. In a normal adult, bladder capacity ranges from 500 to 600 ml.

2. The left ureter is slightly longer than the right because the:
    A. left kidney is higher than the right.
    B. right kidney is higher than the left.
    C. left kidney performs more functions.
    D. left ureter has a three-layered wall.

Answer: A. The left kidney is slightly higher than the right kidney. Therefore, the left ureter needs to be longer to reach the bladder.

3. Urination results from an involuntary and voluntary process. This process is called the:
   A. kidney process.
   B. glomerular filtration rate.
   C. prostate reflex.
   D. micturition reflex.

Answer: D. The micturition reflex is the signal system that occurs when urine fills the bladder. Parasympathetic nerve fibres in the bladder wall cause the bladder to contract and the internal sphincter to relax, which is followed by a voluntary relaxation of the external sphincter.

4. A person on a new health regimen has begun to drink at least 240 ml of water six to eight times per day. The kidneys react to this change by:
   A. producing aldosterone.
   B. secreting renin.
   C. increasing urine output.
   D. secreting erythropoietin.

Answer: C. Urine output typically varies with fluid intake and climate. After ingestion of a large volume of fluid, urine output increases as the body rapidly excretes excess water.

Scoring

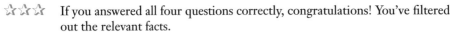 If you answered all four questions correctly, congratulations! You've filtered out the relevant facts.

If you answered three questions correctly, not bad! You've shown you can cope under increased pressure.

If you answered fewer than three questions correctly, read the chapter again and see what you can reabsorb!

13 Fluids, electrolytes, acids and bases

Just the facts

In this chapter, you'll learn:

♦ the way in which fluids are distributed throughout the body
♦ the kidneys' role in electrolyte balance
♦ the body's way of compensating for acid-base imbalances
♦ major acid-base imbalances.

Fluid balance

The health and *homeostasis* (equilibrium of the various body functions) of the human body depend on *fluid, electrolyte* and *acid-base balance*. Factors that disrupt this balance, such as surgery, illness and injury, can lead to potentially fatal changes in metabolic activity. (See *How the body gains and loses fluids*, page 222.)

Featured fluids

Body fluid is made up of water-containing *solutes*, or dissolved substances, that are necessary for physiological functioning. Solutes include electrolytes, glucose, amino acids and other nutrients. The body holds fluid in two basic compartments:

• *Intracellular fluid (ICF)*, found within the individual cells of the body.
• *Extracellular fluid (ECF)*, found in the spaces between cells; includes intravascular fluid (IVF) and interstitial fluid (ISF).

Capillary walls and cell membranes separate the intracellular and extracellular compartments. (See *Fluid compartments*, page 222.)

Fluids, electrolytes, acids, and bases—it's a balancing act to keep the body functioning correctly.

Body shop

How the body gains and loses fluids

Each day, the body gains and loses fluid through several different processes. Typical gains and losses are illustrated below. Gastric, intestinal, pancreatic and biliary secretions are almost completely resorbed and are not usually considered in daily fluid balance.

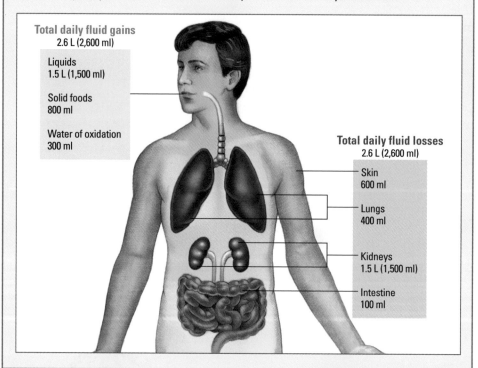

Total daily fluid gains
2.6 L (2,600 ml)

Liquids
1.5 L (1,500 ml)

Solid foods
800 ml

Water of oxidation
300 ml

Total daily fluid losses
2.6 L (2,600 ml)

Skin
600 ml

Lungs
400 ml

Kidneys
1.5 L (1,500 ml)

Intestine
100 ml

Fluid compartments

This illustration shows the two primary fluid compartments in the body: intracellular and extracellular. The extracellular compartment is further divided into the interstitial and intravascular compartments. The interstitial fluid surrounds cells and the intravascular fluid is the fluid portion of blood (plasma). Capillary walls and cell membranes separate intracellular fluids from extracellular fluids. To maintain proper fluid balance, the distribution of fluid within these two compartments must remain relatively constant.

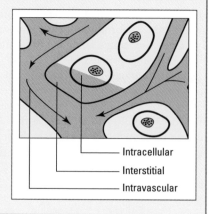

Intracellular

Interstitial

Intravascular

Water weight

Water in the body exists in two major compartments that are separated by capillary walls and cell membranes. About two-thirds of the body's water is found within cells as intracellular fluid (ICF); the other third remains outside cells as extracellular fluid (ECF).

Body weight 70 kg

ICF
ICF accounts for approximately 40% of body weight, or approximately 28 L.

ECF
ECF accounts for approximately 20% of body weight, or approximately 14 L, and includes interstitial fluid and intravascular fluid.

- Interstitial fluid = about for 15% of body weight, or approximately 10.5 L.

- Intravascular fluid (plasma) = about for 5% of body weight, or approximately 3.5 L.

Playing the percentages

Body water contributes to approximately 60% of total body weight in a healthy adult. ICF and ECF comprise about 40% and 20%, respectively, of an adult's total body weight. (See *Water weight*.)

Fluid forms and movement

Fluids in the body generally aren't found in pure forms. They're most commonly found in three different types of solutions: isotonic, hypotonic and hypertonic. The terms are relative and are used in relation to the concentration of ions in fluids relative to the concentration of ions in cells. *Tonicity* is the ability to influence water movement in or out of cells.

No shifting needed

An isotonic solution has the same solute concentration as another solution. For example, normal saline solution is considered isotonic with blood because the concentrations of sodium in the two solutions are approximately equal.

As a result, two equally concentrated fluids in adjacent compartments are already in balance so the fluid inside each compartment stays put; no imbalance means no net fluid shift. Cells won't shrink or swell because there's no gain or loss of water in the cell.

Go low for hypo . . .

A hypotonic solution has a lower solute concentration than another solution. For instance, when one solution contains less sodium than another solution, the first solution is hypotonic compared with the second. As a result, fluid from the first solution—the hypotonic solution—would shift into the second solution until the two solutions had equal concentrations. (Remember, the body constantly strives to maintain a state of balance, or equilibrium.) Administration of a hypotonic solution would cause water to move into the cells, making them swell.

. . . and high for hyper

A hypertonic solution has a higher solute concentration than another solution. For instance, when one solution contains a large amount of sodium and a second solution contains hardly any, the first solution is hypertonic compared with the second solution. As a result, fluid would be drawn from the second solution into the first solution—the hypertonic solution—until the two solutions had equal concentrations. Again, the body strives to maintain a state of equilibrium; therefore, administration of a hypertonic solution would cause water to be drawn out of the cells, making them shrink.

Fluid movement within the cells

Fluids and solutes move constantly within the body. That movement allows the body to maintain *homeostasis*, the constant state of balance the body seeks.

Solutes within the various compartments of the body (intracellular, interstitial and intravascular) move through the membranes that separate those compartments. The membranes are selectively permeable, meaning that they allow some solutes to pass through but not others. Solutes move through membranes at the cellular level by diffusion (movement of particles from an area of high concentration to an area of lower concentration), active transport or osmosis.

Just passing through

Osmosis refers to the passive movement of fluid across a membrane from an area of lower solute concentration and comparatively more fluid into an area of higher solute concentration and comparatively less fluid. Osmosis stops when enough fluid has moved through the membrane to equalise the solute concentration on both sides of the membrane.

Memory jogger

To help remember which fluid belongs to which compartment, keep in mind that *inter* means *between* (as in 'interval') and *intra* means *within* or *inside* (as in 'intravenous').

Needs some effort . . .

In *active transport*, solutes move from an area of lower concentration to an area of higher concentration. Active transport requires energy to make it happen.

. . . and that costs!

The energy required for a solute to move against a concentration gradient comes from a substance called *adenosine triphosphate (ATP)*. Stored in all cells, ATP supplies energy for solute movement in and out of cells. In the process of supplying energy ATP is split into *adenosine diphosphate (ADP)* or *adenosine monophosphate (AMP)*. To maintain the energy supply, ATP must be continually reformed from ADP and AMP using energy produced in breakdown of glucose, fats and proteins.

Some solutes, such as sodium and potassium, use ATP to move in and out of cells in a form of active transport called the *sodium-potassium pump*. Other solutes that require active transport to cross cell membranes include calcium ions, hydrogen ions, amino acids and certain sugars.

In with the good

Water normally enters the body from the GI tract. Each day, the body obtains about 1.5 L of water from consumed liquids and approximately 800 ml more from solid foods, which may consist of up to 97% water. Oxidation of food in the body yields carbon dioxide (CO_2) and about 300 ml of water (water of oxidation).

Out with the bad

Water leaves the body through the skin (in perspiration), lungs (in expired air), GI tract (in faeces) and urinary tract (in urine).

The major pipeline

The main route of water loss is urine excretion with average output of 800 to 2,100 ml over 24 hours. Water losses through the skin (600 ml) and lungs (400 ml) amount to 1 L daily but may increase markedly with strenuous exertion, which predisposes a person to dehydration.

Don't interrupt

In a healthy body, fluid gains match fluid losses to maintain normal physiological functions. However, interruption or dysfunction of one or both of the mechanisms that regulate fluid balance—thirst and the *countercurrent mechanism*—can lead to a fluid imbalance.

I'm parched!

Thirst—the conscious desire for water—is the primary regulator of fluid intake. When the body becomes dehydrated, ECF volume is reduced, causing an increase in sodium concentration and osmolarity.

Thirst—the conscious desire for water—is the primary regulator of fluid intake.

When the sodium concentration reaches about 2 mmol/L above normal, neurones of the thirst centre in the hypothalamus are stimulated. The brain then directs motor neurones to satisfy thirst, causing the individual to drink enough fluid to restore ECF volume.

What comes in must go out

Through the countercurrent mechanism, the kidneys regulate fluid output by modifying urine concentration—that is, by excreting urine of greater or lesser concentration, depending on fluid balance.

Electrolyte balance

Electrolytes are substances that *dissociate* (break up) into electrically charged particles, called *ions*, when dissolved in water. Adequate amounts of each major electrolyte and a proper balance of electrolytes are required to maintain normal physiological functions.

All charged up

Ions may be positively charged (*cations*) or negatively charged (*anions*). Major cations include sodium (Na^+), potassium (K^+), calcium (Ca^{2+}) and magnesium (Mg^{2+}). Major anions include chloride, bicarbonate (HCO_3^-) and phosphate (PO_4^{3-}).

Normally, the electrical charges of cations balance the electrical charges of anions, keeping body fluids electrically neutral.

Shh! We're concentrating

Ions are usually present in low concentrations in body fluids and are expressed in millimoles per litre (mmol/L). ICF and ECF cells are permeable to different substances; therefore, these compartments normally have different electrolyte compositions. (See *Electrolyte composition in ICF and ECF*.)

A delicate balance

Electrolytes profoundly affect the body's water distribution, osmolarity and acid-base balance. Numerous mechanisms within the body help maintain electrolyte balance. Dysfunction or interruption of any of these mechanisms can produce an electrolyte imbalance.

Here are the regulatory mechanisms for common electrolytes:
• The kidneys and a hormone called *aldosterone* are the chief sodium regulators. The small intestine absorbs sodium readily from food, and the skin and kidneys excrete sodium. (See *Osmotic regulation of sodium and water*.)

Electrolyte composition in ICF and ECF

This table shows the electrolyte compositions of intracellular fluid (ICF) and extracellular fluid (ECF). The reference ranges are typical of those reported by NHS laboratories.

Electrolyte	ICF	ECF
Sodium (Na^+)	10 mmol/L	135 to 145 mmol/L
Potassium (K^+)	140 mmol/L	3.5 to 5 mmol/L
Calcium (Ca^{2+})	2.5 mmol/L	1.1 to 1.4 mmol/L (ionised)
Magnesium (Mg^{2+})	20 mmol/L	0.75 to 1.1 mmol/L
Chloride (Cl^-)	4 mmol/L	98 to 108 mmol/L
Bicarbonate (HCO_3^-)	10 mmol/L	24 to 28 mmol/L
Phosphate (PO_4^{3-})	100 mmol/L	0.7 to 1.4 mmol/L

Now I get it!

Osmotic regulation of sodium and water

This flowchart illustrates two compensatory mechanisms used to restore sodium and water balance.

Serum sodium level decreases (water excess).	Serum sodium increases (water deficit).
Serum osmolality drops to less than 280 mOsm/kg.	Serum osmolality increases to more than 300 mOsm/kg.
Thirst decreases, leading to diminished water intake.	Thirst increases, leading to greater water intake.
Antidiuretic hormone (ADH) release is suppressed.	ADH release increases.
Renal water excretion increases.	Renal water excretion decreases.

Serum osmolality normalises.

- The kidneys also regulate potassium through aldosterone action. Most potassium is absorbed from food in the GI tract; normally, the amount excreted in urine equals dietary potassium intake.
- Calcium in the blood is typically in equilibrium with calcium salts in bone. Parathyroid hormone (PTH) is the main regulator of calcium, controlling both calcium uptake from the GI tract and calcium excretion by the kidneys.
- Magnesium is governed by aldosterone, which controls renal magnesium reabsorption. Absorbed from the GI tract, magnesium is excreted in urine, breast milk and saliva.
- The kidneys also regulate chloride. Chloride ions move in conjunction with sodium ions.
- The kidneys regulate bicarbonate, excreting, absorbing, or forming it. Bicarbonate, in turn, plays a vital part in acid-base balance.
- The kidneys regulate phosphate. Absorbed from food, phosphate is incorporated with calcium in bone. PTH governs calcium and phosphate levels.

You're an electrolyte superstar—helping to regulate potassium, chloride, bicarbonate and phosphate.

Acid-base balance

Physiological survival requires *acid-base balance*, a stable concentration of hydrogen ions in body fluids.

An acid remark

An *acid* is a substance that yields hydrogen ions when it *dissociates* (changed from a complex to a simpler compound) in solution. A strong acid dissociates almost completely, releasing a large number of hydrogen ions.

A base reply

A *base* dissociates in water, releasing ions that can combine with hydrogen ions. Like a strong acid, a strong base dissociates almost completely, releasing many ions.

The hydrogen ion concentration of a fluid determines whether it's *acidic* or *basic* (alkaline). A *neutral* solution, such as pure water, dissociates only slightly. (See *Understanding pH*.)

Keep those ions coming

The body produces acids, thus yielding hydrogen ions, through the following mechanisms:
- Protein catabolism yields nonvolatile acids, such as sulphuric, phosphoric and uric acids.
- Fat oxidation produces acid ketone bodies.

Will a fluid be acidic or basic? Only the hydrogen ion concentration knows for sure.

Now I get it!

Understanding pH

Hydrogen ion (H^+) concentration is commonly expressed as pH, which indicates the degree of acidity or alkalinity of a solution:
- A pH of 7 indicates neutrality or equal amounts of H^+ and hydroxyl ions (OH^-).
- An acidic solution contains more H^+ ions than OH^- ions; its pH is less than 7.
- An alkaline solution contains fewer H^+ ions than OH^- ions; its pH exceeds 7.
- Overall, as H^+ ion concentration increases, pH goes down.

1 means 10

Because pH is an exponential expression, a change of one pH unit reflects a 10-fold difference in actual H^+ concentration. For instance, a solution with a pH of 7 has 10 times more H^+ than a solution with a pH of 8.

- Anaerobic glucose catabolism produces lactic acid.
- Intracellular metabolism yields CO_2 as a by-product; CO_2 dissolves in body fluids to form carbonic acid (H_2CO_3).

Balancing buffers

Normally, even with the production of these acids, the body's pH control mechanism is so effective that arterial blood pH stays within a narrow range: 7.35 to 7.45. This acid-base balance is maintained by buffer systems and the lungs and kidneys, which neutralise and eliminate acids as rapidly as they are formed.

Dysfunction or interruption of a buffer system or other governing mechanism can cause an acid-base imbalance. *Acidosis* occurs when the hydrogen ion concentration increases above normal (or the bicarbonate concentration falls). *Alkalosis* occurs when the hydrogen ion concentration falls below normal (or the bicarbonate level is increased). Acidosis or alkalosis can occur in the body as a result of respiratory or metabolic disorders, or a combination of both. (See *Understanding respiratory and metabolic alkalosis and acidosis*, page 230.)

(Text continues on page 234.)

Now I get it!

Understanding respiratory and metabolic alkalosis and acidosis

What happens in respiratory alkalosis

Step 1

When pulmonary ventilation increases above the amount needed to maintain normal carbon dioxide (CO_2) levels, excessive amounts of CO_2 are exhaled. This causes hypocapnia (a fall in partial pressure of arterial carbon dioxide [$Paco_2$]), which leads to a reduction in carbonic acid (H_2CO_3) production, a loss of hydrogen (H^+) and bicarbonate ions (HCO_3^-), and a subsequent rise in pH. Look for a pH level above 7.45, a $Paco_2$ level below 4.7 kPa, and an HCO_3^- level below 22 mmol/L (as shown at right).

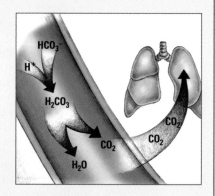

Step 2

In defence against the rising pH, H^+ are pulled out of the cells and into the blood in exchange for potassium ions (K^+). The H^+ entering the blood combines with HCO_3^- to form H_2CO_3, which lowers the pH. Look for a further decrease in HCO_3^- levels, a fall in pH, and a fall in serum K^+ levels (hypokalaemia).

Step 3

Hypocapnia stimulates the carotid and aortic bodies and the medulla, which causes an increase in heart rate without an increase in blood pressure (as shown at right). Look for angina, electrocardiogram changes, restlessness and anxiety.

Step 4

Simultaneously, hypocapnia produces cerebral vasoconstriction, which prompts a reduction in cerebral blood flow. Hypocapnia also overexcites the medulla, pons and other parts of the autonomic nervous system. Look for increasing anxiety, diaphoresis, dyspnoea, alternating periods of apnoea and hyperventilation, dizziness and tingling in the fingers or toes.

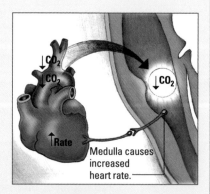

Step 5

When hypocapnia lasts more than 6 hours, the kidneys increase secretion of HCO_3^- and reduce excretion of H^+ (as shown on the right). Periods of apnoea may result if the pH remains high and the $Paco_2$ remains low. Look for slowing of the respiratory rate, hypoventilation and Cheyne-Stokes respirations (altered breathing pattern).

Step 6

Continued low $Paco_2$ increases cerebral and peripheral hypoxia from vasoconstriction. Severe alkalosis inhibits calcium ionisation (Ca^{2+} formation) which, in turn, causes increased nerve excitability and muscle contractions. Eventually, the alkalosis overwhelms the central nervous system (CNS) and the heart. Look for decreasing level of consciousness (LOC), hyper-reflexia, carpopedal spasm, tetany, arrhythmias, seizures and coma.

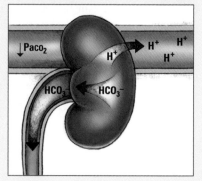

Understanding respiratory and metabolic alkalosis and acidosis (Continued)

What happens in respiratory acidosis

Step 1

When pulmonary ventilation decreases, retained CO_2 combines with water (H_2O) to form H_2CO_3 in larger-than-normal amounts. The H_2CO_3 dissociates to release free H^+ and HCO_3^-. The excessive H_2CO_3 causes a drop in pH. Look for a $Paco_2$ level above 6 kPa and a pH level below 7.35.

Step 2

As the pH level falls, 2,3-diphosphoglycerate (2,3-DPG) increases in the red blood cells and causes a change in haemoglobin (Hb) that makes the Hb release oxygen (O_2). The altered Hb, now strongly alkaline, picks up H^+ and CO_2, thus eliminating some of the free H^+ and excess CO_2 (as shown on the right). Look for decreased arterial oxygen saturation.

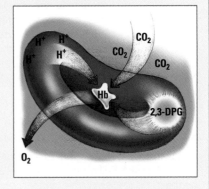

Step 3

Whenever $Paco_2$ increases, CO_2 builds up in all tissues and fluids, including cerebrospinal fluid and the respiratory centre in the medulla. The CO_2 reacts with H_2O to form H_2CO_3, which then breaks into free H^+ and HCO_3^-. The increased amount of CO_2 and H^+ stimulate the respiratory centre to increase the respiratory rate. An increased respiratory rate expels more CO_2 and helps to reduce the CO_2 level in the blood and other tissues. Look for rapid, shallow respirations and a decreasing $Paco_2$.

Step 4

Eventually, CO_2 and H^+ cause cerebral blood vessels to dilate, which increases blood flow to the brain. That increased flow can cause cerebral oedema and depress CNS activity. Look for headache, confusion, lethargy, nausea, or vomiting. As respiratory mechanisms fail, the increasing $Paco_2$ stimulates the kidneys to retain HCO_3^- and sodium (Na^+) and to excrete H^+, some of which is excreted in the form of ammonium (NH_4^+). The additional HCO_3^- and Na^+ combine to form extra sodium bicarbonate ($NaHCO_3$), which is then able to buffer more free H^+ ions (as shown at right). Look for increased acid content in the urine, increasing serum pH and HCO_3^- levels, and shallow, depressed respirations.

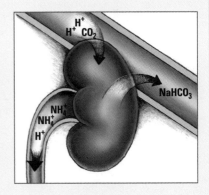

Step 5

As the concentration of H^+ overwhelms the body's compensatory mechanisms, the H^+ move into the cells and K^+ move out. A concurrent lack of O_2 causes an increase in the anaerobic production of lactic acid, which further skews the acid-base balance and critically depresses neurological and cardiac functions. Look for hyperkalaemia, arrhythmias, increased $Paco_2$, decreased partial pressure of arterial oxygen, decreased pH and decreased LOC.

(continued)

Understanding respiratory and metabolic alkalosis and acidosis (Continued)

What happens in metabolic alkalosis

Step 1

As HCO_3^- starts to accumulate in the body, chemical buffers (in extracellular fluid [ECF] and cells) bind with them. No signs are detectable at this stage.

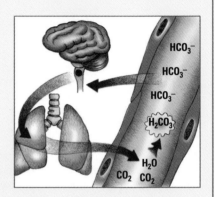

Step 2

Excess HCO_3^- that fail to bind with chemical buffers elevate serum pH levels, which, in turn, depress chemoreceptors in the medulla. Depression of those chemoreceptors causes a decrease in respiratory rate, which increases the Pa_{CO_2}. The additional CO_2 combines with H_2O to form H_2CO_3 (as shown at right). Note: Lowered O_2 levels limit respiratory compensation. Look for a serum pH level above 7.45, an HCO_3^- level above 26 mmol/L, a rising Pa_{CO_2}, and slow, shallow respirations.

Step 3

When the HCO_3^- level exceeds 28 mmol/L, the renal glomeruli can no longer reabsorb excess HCO_3^-. That excess HCO_3^- is excreted in the urine; H^+ are retained. Look for alkaline urine and pH and HCO_3^- levels that slowly return to normal.

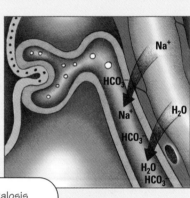

Step 4

To maintain electrochemical balance, the kidneys excrete excess Na^+, H_2O and HCO_3^- (as shown at right). Look for polyuria initially, and then signs of hypovolaemia, including thirst and dry mucous membranes.

Step 5

Lowered H^+ levels in the ECF cause the ions to diffuse out of the cells. To maintain the balance of charge across the cell membrane, extracellular K^+ move into the cells. Look for signs of hypokalaemia: anorexia, muscle weakness, loss of reflexes and others.

Metabolic alkalosis ultimately progresses to tetany, aggression, irritability, disorientation and seizures.

Step 6

As H^+ levels decline, the level of Ca^{2+} ionisation decreases. That decrease in ionisation makes nerve cells more permeable to Na^+. The increase in Na^+ moving into nerve cells stimulate neural impulses and produce overexcitability of the peripheral nervous system and CNS. Look for tetany, aggression, irritability, disorientation and seizures.

Understanding respiratory and metabolic alkalosis and acidosis (Continued)

What happens in metabolic acidosis

Step 1

As H⁺ start to accumulate in the body, chemical buffers (plasma HCO_3^- and proteins) in the cells and ECF bind with them (as shown at right). No signs are detectable at this stage.

Step 2

Excess H⁺ (which can't bind with the buffers) decrease the pH and stimulate chemoreceptors in the medulla to increase the respiratory rate. The increased respiratory rate lowers the Pa_{CO_2}, which allows more H⁺ to bind with HCO_3^-. Respiratory compensation occurs within minutes, but isn't sufficient to correct the imbalance (see middle illustration). Look for a pH level below 7.35, an HCO_3^- level below 22 mmol/L, a decreasing Pa_{CO_2} level, and rapid, deeper respirations.

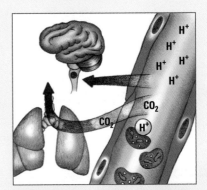

Step 3

Healthy kidneys try to compensate for the acidosis by secreting excess H⁺ into the renal tubules. Those ions are buffered by phosphate or ammonia and then are excreted into the urine in the form of a weak acid. Look for acidic urine.

Step 4

Each time a H⁺ is secreted into the renal tubules, a Na^+ and an HCO_3^- are absorbed from the tubules and returned to the blood. Look for pH and HCO_3^- levels that slowly return to normal.

Step 5

Excess H⁺ in the ECF diffuse into cells. To maintain the balance of the charge across the membrane, the cells release K^+ into the blood (as shown at right). Look for signs of hyperkalaemia, including colic and diarrhoea, weakness or flaccid paralysis, tingling and numbness in the extremities, bradycardia, a tall T wave, a prolonged PR interval, and a wide QRS complex.

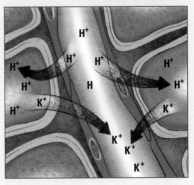

Step 6

Excess H⁺ alter the normal balance of K^+, Na^+ and Ca^{2+}, leading to reduced excitability of nerve cells. Look for signs and symptoms of progressive CNS depression, including lethargy, dull headache, confusion, stupor and coma.

Buffer systems

Buffers are substances that prevent changes in the pH by binding or releasing hydrogen ions. *Buffer systems* reduce the effect of an abrupt change in hydrogen ion concentration by converting a strong acid or base (which normally would dissociate completely) into a weak acid or base (which releases fewer hydrogen ions).

Buffer systems that help maintain acid-base balance include:
- sodium bicarbonate–carbonic acid
- phosphate
- protein.

> We team up to keep blood pH normal, between 7.35 and 7.45.

One from the kidneys, one from the lungs

The *sodium bicarbonate–carbonic acid* buffer system is the major buffer in ECF. Sodium bicarbonate concentration is regulated by the kidneys, and carbonic acid concentration is regulated by the lungs. Both components of this buffer are replenished continually. As a result of the buffering action, the strong base (sodium hydroxide) is replaced by sodium bicarbonate and water (H_2O). Sodium hydroxide dissociates almost completely and releases large amounts of hydroxyl. If a strong acid is added, the opposite occurs.

Finesse with phosphate

A *phosphate* buffer system works by regulating the pH of fluids as they pass through the kidneys. It's also the main intracellular buffer.

Absorbing ions

Protein buffers can exist in the form of acids or alkaline salts. In the *protein* buffer system, intracellular proteins absorb hydrogen ions generated by the body's metabolic processes and may release excess hydrogen as needed.

Lungs

The carbonic acid—bicarbonate buffer system changes the pH of the blood in 3 minutes or less by changing the breathing rate. A decreased respiratory rate decreases the release of CO_2; this causes an increase in the amount of carbonic acid and free hydrogen ions in the blood resulting in a pH rise. Present in all acids, hydrogen ions are protons that can be added to or removed from a solution to change the pH.

Respiration plays a crucial role in controlling pH. The lungs excrete CO_2 and regulate the carbonic acid content of the blood. Carbonic acid is derived from the CO_2 and water that are released as by-products of cellular metabolic activity.

Stick to the formula

CO_2 is soluble in blood plasma. Some of the dissolved gas reacts with water to form carbonic acid, a weak acid that partially breaks apart to form hydrogen and bicarbonate ions. These three substances are in equilibrium, as reflected in the following formula:

$$CO_2 + H_2O \leftrightarrow H_2CO_3 \leftrightarrow H^+ + HCO_3^-.$$

CO_2 dissolved in plasma is in equilibrium with CO_2 in the lung alveoli (expressed as a partial pressure [Pco_2]). Thus, an equilibrium exists between alveolar Pco_2 and the various forms of CO_2 present in the plasma, as expressed by the following formula:

$$Pco_2 \leftrightarrow CO_2 + H_2O \leftrightarrow H_2CO_3 \leftrightarrow H^+ + HCO_3^-.$$

Don't hold your breath!

A change in the rate or depth of respirations can alter the CO_2 content of alveolar air and the alveolar Pco_2. A change in alveolar Pco_2 produces a corresponding change in the amount of carbonic acid formed by dissolved CO_2. In turn, these changes stimulate the respiratory centre to modify respiratory rate and depth.

An increase in alveolar Pco_2 raises the blood concentration of CO_2 and carbonic acid. This, in turn, stimulates the respiratory centre to increase respiratory rate and depth. As a result, alveolar Pco_2 decreases, which leads to a corresponding drop in the carbonic acid and CO_2 concentrations in blood. (See *How respiratory mechanisms affect blood pH*, page 236.)

A decrease in respiratory rate and depth has the reverse effect; it raises alveolar Pco_2 which, in turn, triggers an increase in the blood's CO_2 and carbonic acid concentrations.

Kidneys

In addition to excreting various acid waste products, the kidneys help manage acid-base balance by regulating the blood's bicarbonate concentration. They do so by permitting bicarbonate reabsorption from tubular filtrate and by forming additional bicarbonate to replace that used in buffering acids.

Now I get it!

How respiratory mechanisms affect blood pH

A decrease in blood pH stimulates the respiratory centre, causing *hyperventilation*. As a result, carbon dioxide (CO_2) levels decrease and, therefore, less carbonic acid and fewer hydrogen ions remain in the blood. Consequently, blood pH increases, possibly reaching a normal level.

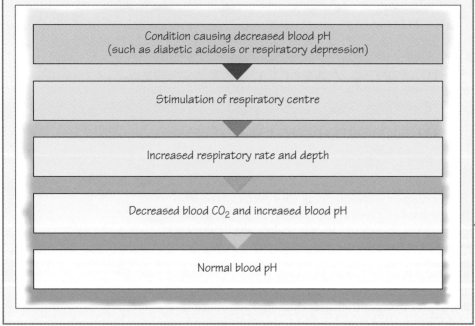

Condition causing decreased blood pH (such as diabetic acidosis or respiratory depression)

Stimulation of respiratory centre

Increased respiratory rate and depth

Decreased blood CO_2 and increased blood pH

Normal blood pH

A rise in carbon dioxide or a decrease in pH in the blood causes hyperventilation. That means we have to work harder!

Renal tubular ion secretion

Recovery and formation of bicarbonate in the kidneys depend on hydrogen ion secretion by the renal tubules in exchange for sodium ions. Sodium ions are then simultaneously reabsorbed into the circulation from the tubular filtrate.

Influential enzyme

The enzyme *carbonic anhydrase* influences tubular epithelial cells to form carbonic acid from CO_2 and water. Carbonic acid quickly dissociates into both hydrogen and bicarbonate ions. Hydrogen ions (H^+) enter the tubular filtrate in exchange for sodium ions; bicarbonate ions enter the bloodstream along with the sodium ions that have been absorbed from the filtrate. Bicarbonate is then reabsorbed from the tubular filtrate.

Ions hanging out together

Each hydrogen ion secreted into the tubular filtrate joins with a bicarbonate ion to form carbonic acid, which rapidly dissociates into CO_2 and water. The CO_2 diffuses into the tubular epithelial cells, where it can combine with more water and lead to the formation of more carbonic acid.

Bicarbonate reabsorption

The remaining water molecule in the tubular filtrate is eliminated in the urine. As each hydrogen ion enters the tubular filtrate to combine with a bicarbonate ion, a bicarbonate ion in the tubular epithelial cells diffuses into the circulation. This process is termed bicarbonate reabsorption. (However, the bicarbonate ion that enters the circulation isn't the same one as in the tubular filtrate.)

Formation of ammonia and phosphate salts

To form more bicarbonate, the kidneys must secrete additional hydrogen ions in exchange for sodium ions. For the renal tubules to continue secreting hydrogen ions, the excess ions must combine with other substances in the filtrate and be excreted. Excess hydrogen ions in the filtrate may combine with *ammonia* (NH_3), which is produced by the renal tubules, or with phosphate salts present in the tubular filtrate.

More ions on the move

After diffusing into the filtrate, ammonia joins with the secreted hydrogen ions, forming ammonium ions. These ions are excreted in the urine with chloride and other anions; each secreted ammonia molecule eliminates one hydrogen ion in the filtrate.

At the same time, sodium ions that have been absorbed from the filtrate and exchanged for hydrogen ions enter the circulation, as does the bicarbonate formed in the tubular epithelial cells. Some secreted hydrogen ions combine with a *disodium hydrogen phosphate* (Na_2HPO_4) in the tubular filtrate. Each of the secreted hydrogen ions that joins with the disodium salt changes to the monosodium salt sodium dihydrogen phosphate (NaH_2PO_4). The sodium ion released in this reaction is absorbed into the circulation along with a newly formed bicarbonate ion.

Factors affecting bicarbonate formation

The rate of bicarbonate formation by renal tubular epithelial cells is affected by two factors:
- the amount of dissolved CO_2 in the plasma
- the potassium content of the tubular cells.

If the plasma CO_2 level rises, renal tubular cells form more bicarbonate. Increased plasma CO_2 encourages greater carbonic acid formation by the renal tubular cells.

Chain reaction

Partial dissociation of carbonic acid results in more hydrogen ions for excretion into the tubular filtrate and additional bicarbonate ions for entry into the circulation. This, in turn, increases the plasma bicarbonate level and reduces the plasma level of dissolved CO_2 towards normal. If the plasma CO_2 level decreases, renal tubular cells form less carbonic acid.

Because fewer hydrogen ions are formed and excreted, fewer bicarbonate ions enter the circulation. The plasma bicarbonate level then falls accordingly.

Special K

The potassium content of renal tubular cells also helps regulate plasma bicarbonate concentration by affecting the rate at which the renal tubules secrete hydrogen ions. Tubular cell potassium content and hydrogen ion secretion are interrelated; potassium and hydrogen ions are secreted at rates that vary inversely. Hydrogen ion secretion increases if tubular secretion of potassium ions falls; hydrogen secretion declines if tubular secretion of potassium ions increases.

For each hydrogen ion secreted into the tubular filtrate, an additional bicarbonate ion enters the blood plasma. Consequently, increased tubular secretion of hydrogen ions leads to a rise in the plasma bicarbonate content.

Depletion of body potassium causes more bicarbonate to enter the circulation; the plasma bicarbonate level then rises above normal. When the body contains excess potassium, the tubules secrete more potassium. As a result, fewer hydrogen ions are secreted, less bicarbonate forms and the plasma bicarbonate concentration decreases.

Potassium secretion by tubular epithelial cells decreases and hydrogen ion secretion rises in patients with potassium depletion from vomiting or diarrhoea.

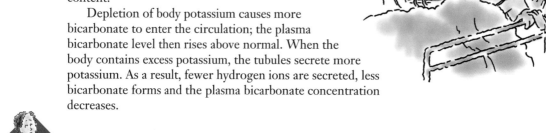

Quick quiz

1. A solution with a pH of less than 7 is considered:
   A. acidic.
   B. alkaline.
   C. solute.
   D. hypotonic.

   Answer: A. A solution with a pH of less than 7 contains more hydrogen ions than hydroxyl ions and is considered acidic.

2. The body compensates for chronic respiratory alkalosis by developing:
    A. metabolic alkalosis.
    B. respiratory acidosis.
    C. metabolic acidosis.
    D. a phosphate buffer system.

Answer: C. The body compensates for chronic respiratory alkalosis by developing metabolic acidosis.

3. The two factors that affect the rate of bicarbonate formation by renal tubular epithelial cells are:
    A. amount of aldosterone in the system and urine production.
    B. amount of dissolved CO_2 in the plasma and the potassium content of the tubular cells.
    C. amount of dissolved potassium in the plasma and the CO_2 content of the tubular cells.
    D. amount of ammonia produced by the renal tubules and the phosphate salts present in the tubular filtrate.

Answer: B. The rate of bicarbonate formation by renal tubular epithelial cells is affected by the amount of dissolved CO_2 in the plasma and the potassium content of tubular cells.

Scoring

☆☆☆ If you answered all three questions correctly, give yourself a pat on the back! Your knowledge is well balanced.

☆☆ If you answered two questions correctly, good for you! You're benefiting from all the right buffer systems.

☆ If you answered only one question correctly, check out this chapter again. It may take time to find your equilibrium.

Just the facts

In this chapter, you'll learn:

♦ basic functions of the skin

♦ skin layers and their components

♦ the accessory structures (hair, nails and glands) of the integumentary system.

A look at the integumentary system

The integumentary system is the largest body system and includes the skin, or *integument*, and its accessory structures (the hair, nails and certain glands).

Not just another pretty face

The integumentary system performs a number of vital functions, including:

• protection of inner body structures
• sensory perception
• regulation of body temperature
• excretion of some body fluids and waste products.

Protection

The skin maintains the integrity of the body surface through the upward migration of the lower layers and shedding. It can repair surface wounds by intensifying normal cell replacement mechanisms. The skin's top layer, known as the *epidermis*, protects the body against noxious chemicals and invasion from pathogens.

The skin serves as the body's primary defence mechanism, protecting the body from invaders.

Langerhans' cells to the rescue

Langerhans' cells are specialised cells within the epidermis. They enhance the body's immune response through phagocytosis and help activate other white cells in the body.

The skin's own sun block

Melanocytes, another type of skin cell, protect the skin by producing the brown pigment *melanin*, which provides some protection against ultraviolet (UV) light (irradiation). Exposure to UV light can stimulate melanin production.

> Melanocytes help protect the body from ultraviolet light . . . but that doesn't mean I can do without my sunshade!

Sensory perception

Sensory nerve fibres originate in the nerve roots along the spine and supply sensation to specific areas of the skin known as *dermatomes*.

Just sensational

These nerve fibres transmit various sensations, such as temperature, touch, pressure, pain and itching, from the skin to the central nervous system. Autonomic nerve fibres carry impulses to smooth muscle in the walls of the skin's blood vessels, to the muscles around the hair roots and to the sweat glands.

Body temperature regulation

Abundant nerves, blood vessels and eccrine glands within the skin's deeper layer, the *dermis*, help control body temperature (thermoregulation).

Warming up . . .

When the skin is exposed to cold or internal body temperature falls, surface blood vessels constrict, decreasing blood flow and thereby conserving body heat.

. . . and cooling down

If the skin becomes too hot or internal body temperature rises, small arteries within the skin dilate, increasing blood flow, which in turn reduces body heat. (See *The skin's role in thermoregulation*, page 242.)

Excretion

The skin is also an excretory organ. The sweat glands excrete sweat, which contains water, electrolytes, urea and lactic acid.

Now I get it!

The skin's role in thermoregulation

Abundant nerves, blood vessels and eccrine glands within the skin's deeper layer aid thermoregulation (control of body temperature). The first part of the flowchart shows how the body conserves body heat. The second part of the flowchart shows how the body reduces body heat. Here's how the skin does its job.

It's time to warm up

The skin becomes exposed to cold or internal body temperature falls.

Blood vessels constrict in response to stimuli from the autonomic nervous system.

Blood flow decreases through the skin and body heat is conserved.

Now let's cool things off

The skin becomes too hot or internal body temperature rises.

Small arteries in the second skin layer dilate (expand).

Increased blood flow reduces body heat. If this doesn't lower temperature, the eccrine glands act to increase sweat production and evaporation cools the skin.

Water works

While it eliminates body wastes through its more than two million pores, the skin also prevents body fluids from escaping. Here, the skin is again protecting the body by preventing dehydration caused by loss of internal body fluids—as well as maintaining these levels by regulating the content and volume of sweat. It also keeps unwanted fluids in the environment from entering the body.

Production site

Vitamin D is produced in the skin after exposure to UV light from the sun or artificial sources (occurs naturally in a small range of foods). It is then carried in the circulation to the liver, where it is converted into the prohormone calcidiol. Circulating calcidiol may then be converted into calcitriol, the biologically active form of vitamin D, which regulates (amongst other things) the concentration of calcium and phosphate in the blood, promoting the healthy mineralisation, growth and remodelling of bone.

Vitamin D deficiency can result in musculoskeletal problems in both children and adults.

Skin layers

Two distinct layers of skin, the *epidermis* and *dermis*, lie above a third layer of *subcutaneous tissue*—sometimes called the *hypodermis*. (See *A close look at skin*, page 244.)

Epidermis

The *epidermis* is the outermost layer and varies in thickness from less than 0.1 mm on the eyelids to more than 1 mm on the palms and soles. It's partially translucent, meaning it allows some light to pass through it.

The ins and outs (and in betweens)

The epidermis is composed of avascular, stratified, squamous (scaly or platelike) epithelial tissue and is divided into five distinct layers. Each layer is named for its structure or function:
• The *stratum corneum* is the outermost non-viable layer and consists of tightly arranged layers of cellular membranes and keratin.
• The *stratum lucidum* blocks water penetration or loss. It is more apparent in the thick skin of the palms or soles.
• The *stratum granulosum* is responsible for keratin formation and, like the stratum lucidum, may be missing in some thin skin.
• The *stratum spinosum* also helps with keratin formation and is rich in ribonucleic acid.
• The *stratum basale* is the innermost layer and produces new cells to replace the superficial keratinised cells that are continuously shed or worn

Zoom in

A close look at skin

Major components of skin include the epidermis, dermis and epidermal appendages.

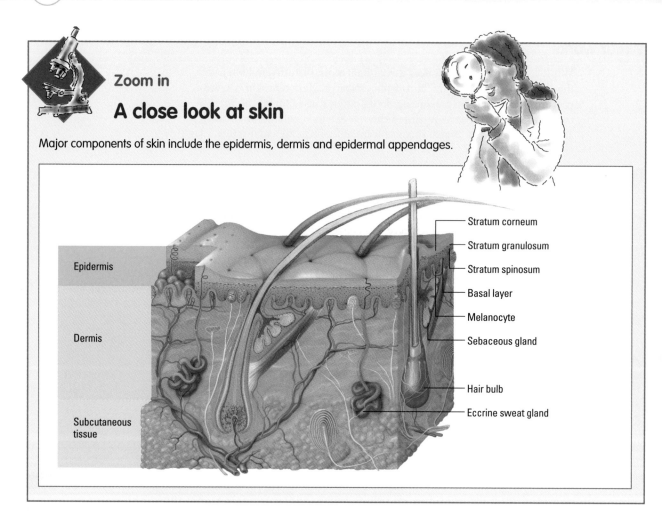

away. The accumulation of keratin within each layer (from the innermost to outermost) is a continual and dynamic process.

No blood, just dermal papillae

The epidermis doesn't contain blood vessels. Food, vitamins and oxygen are transported to this layer through fingerlike structures in the dermis, called *papillae*, which contain a network of tiny blood vessels. These papillae act to increase contact between the dermal layers.

Dermis

The *dermis* is the skin's second layer. It's an elastic tissue layer that contains and supports blood vessels, lymphatic vessels, nerves and most of the accessory structures.

What's in the matrix?

Most of the dermis is made up of extracellular material called *matrix*. Matrix contains:
* *collagen*, a protein made by fibroblasts that gives strength and resilience to the dermis
* *elastic fibres* that bind the collagen and make the skin flexible.

Making a good impression

The dermis itself has two layers:
* The *papillary dermis* has fingerlike projections, *papillae*, that connect the dermis to the epidermis. During early foetal development, epidermal ridges develop as the epidermis conforms to the contours of the underlying papillae. The pattern of these ridges is unique to the individual and help increase grip of the hands (and feet) by increasing friction along the points of contact. Sweat glands open on to the surface of these epidermal ridges causing unique impressions that are left behind when a smooth object is touched. These impressions are commonly known as *fingerprints* and do not normally change throughout life except to enlarge.
* The *reticular dermis* covers a layer of subcutaneous tissue. It's made of collagen fibres and provides strength, structure and elasticity to the skin.

Subcutaneous tissue

Beneath the dermis is the third layer—*subcutaneous tissue*—which is a layer of fat. It contains larger blood vessels and nerves, as well as adipose cells, which are filled with fat. This subcutaneous fat layer lies on the muscles and bones. Functions of subcutaneous tissue include insulation, shock absorption and storage of energy reserves.

Epidermal appendages

Numerous epidermal appendages occur throughout the skin. They include the hair, nails, sebaceous glands and sweat glands. (See *Skin, hair and nail changes with aging*, page 246.)

Hair

Hairs are long, slender shafts composed of the protein *keratin*. At the expanded lower end of each hair is a bulb or root. On its undersurface, the root is indented by a *hair papilla*, a cluster of connective tissue and blood vessels.

Memory jogger

You can keep straight which skin layer is which by remembering that the prefix **epi-** means 'upon'. Therefore, the **epi**dermis is upon, or on top of, the dermis.

When arrector pili muscles contract, hair stands on end.

As time goes by . . .

Skin, hair and nail changes with ageing

In the integumentary system, age-related changes can involve the skin, hair and nails.

The skinny

As people age, their skin changes. For example, they may notice lines around their eyes (crow's feet), mouth and nose. These lines result from subcutaneous fat loss, dermal thinning, decreasing collagen and elastin production and a 50% decline in cell replacement. Women's skin shows signs of aging about 10 years earlier than men's because it's thinner and drier.

Because of the decreased rate of skin cell replacement, wounds may heal more slowly and be prone to infection in older adults. In the extreme elderly, skin loses its elasticity and may seem almost transparent.

Other changes may include a decrease in the size, number, and function of sweat glands. Combined with a loss of subcutaneous fat, this makes the regulation of body temperature more difficult.

Melanocyte production also decreases as a person ages; however, melanocytes often enlarge and proliferate in localised areas, causing brown spots (senile lentigo). This typically occurs in areas regularly exposed to the sun. Other common skin conditions in older people include senile keratosis (dry, harsh skin) and senile angioma (a benign tumour of dilated blood vessels caused by weakened capillary walls).

Hairy situation

Hair changes also occur with aging. Hair pigment decreases and hair may turn grey or white. This loss of pigment makes hair thinner; by age 70, it's baby fine again. Hormonal changes cause pubic hair loss. At the same time, facial hair commonly increases in postmenopausal women and decreases in aging men.

Nailed down

Aging may also alter nails. They may grow at different rates, and longitudinal ridges, flaking, brittleness and malformations may increase. Toenails may also discolour.

It will literally make your hair stand on end

Each hair lies within an epithelium-lined sheath called a *hair follicle*. A bundle of smooth-muscle fibres, *arrector pili*, extends through the dermis to attach to the base of the follicle. When these muscles contract, hair stands on end. Hair follicles also have a rich blood and nerve supply.

Nails

The *nails* are situated over the distal surface of the end of each finger and toe. Nails are composed of a specialised type of keratin.

On a bed of nails

The visible *nail plate*, surrounded on three sides by nail folds, lies on the nail bed. The nail plate is formed by the nail matrix, which extends proximally in the region of 0.5 cm beneath the nail fold (*eponychium*).

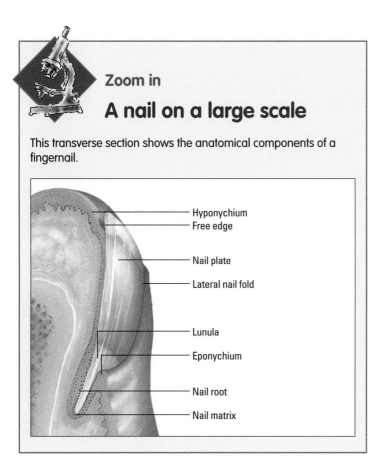

Zoom in

A nail on a large scale

This transverse section shows the anatomical components of a fingernail.

- Hyponychium
- Free edge
- Nail plate
- Lateral nail fold
- Lunula
- Eponychium
- Nail root
- Nail matrix

Landing on the lunula

The distal portion of the matrix shows through the nail as a pale crescent-moon-shaped area. This is called the *lunula*. The translucent nail plate distal to the lunula exposes the nail bed. The vascular bed imparts the characteristic pink appearance under the nails. The *hyponychium* is the epithelium located beneath the nail plate at the junction between the free edge and the skin of the fingertip. It forms a seal that protects the nail bed. (See *A nail on a large scale*.)

Sebaceous glands

Sebaceous glands are part of the hair follicle and occur on all parts of the skin except the palms and soles. They're most prominent on the scalp, face, upper torso and genitalia.

A (small) miracle oil

The sebaceous glands produce *sebum*, a mixture of keratin, fat and cellular debris. Combined with sweat, sebum forms a moist, oily, acidic film that's mildly antibacterial and antifungal and that protects the skin surface. Sebum exits through the hair follicle opening to reach the skin surface.

Sweat glands

There are two types of sweat glands: eccrine glands and apocrine glands.

Eccrine glands

The *eccrine glands* are widely distributed throughout the body and produce an odourless, watery fluid with a sodium concentration equal to that of plasma. A duct from the coiled secretory portion passes through the dermis and epidermis, opening onto the skin surface.

Eccrine glands secrete fluid in response to stress.

Stressed out

Eccrine glands in the palms and soles secrete fluid mainly in response to emotional stress. The remaining three million eccrine glands respond primarily to thermal stress, effectively regulating temperature. Eccrine glands are found everywhere except the lips and glans penis.

Apocrine glands

The *apocrine glands* are located chiefly in the axillary (underarm) and anogenital (groin) areas. They have a coiled secretory portion that lies deeper in the dermis than that of the eccrine glands. A duct connects an apocrine gland to the upper portion of the hair follicle.

Oh no, B.O.

Apocrine glands begin to function at puberty. They have no known biological function, although they are activated by stress. As bacteria decompose the fluids produced by these glands, body odour occurs. However, it should be noted that sweat itself has some antibacterial properties.

Quick quiz

1. The main functions of the skin include:
    A. support, nourishment and sensation.
    B. protection, sensory perception and temperature regulation.
    C. fluid transport, sensory perception and aging regulation.
    D. protection, motor response and filtration.

Answer: B. The skin's main functions involve protection from injury, noxious chemicals and bacterial invasion; sensory perception of touch, temperature and pain; and regulation of body heat.

2. The outermost layer of the skin is the:
   A. epidermis.
   B. dermis.
   C. hypodermis.
   D. papillary dermis.

Answer: A. The outermost layer of the skin, composed of avascular, stratified and squamous epithelial tissue, is the epidermis.

3. Which integumentary system structure is considered an epidermal appendage?
   A. blood vessel
   B. nerve
   C. stratum basale
   D. hair

Answer: D. The appendages of the epidermis are the nails, hair, sebaceous glands, eccrine glands and apocrine glands.

4. Sebum is a mixture of:
   A. cellular debris, fat and keratin.
   B. collagen and elastin.
   C. watery fluid and sodium.
   D. protein, water and electrolytes.

Answer: A. Sebum is produced by the sebaceous glands and is a mixture of keratin, fat and cellular debris.

5. The sweat glands that are widely distributed throughout the body are:
   A. apocrine.
   B. eccrine.
   C. adipose.
   D. sebaceous.

Answer: B. Eccrine glands are widely distributed throughout the body and produce an odourless, watery fluid.

Scoring

☆☆☆ If you answered all five questions correctly, amazing! You've got the integumentary system covered.

☆☆ If you answered four questions correctly, awesome. You're scratching beneath the surface of this body system.

☆ If you answered fewer than four questions correctly, no sweat. Just go back and review the chapter.

Reproductive system  15

Just the facts

In this chapter, you'll learn:

♦ structure and functions of the male and female reproductive systems

♦ male hormone production and its effects on sexual development

♦ female hormone production and its effects on menstruation

♦ the anatomical structure and functions of the female breast.

A look at the reproductive systems

Anatomically, the main distinction between the male and female is the presence of conspicuous external genitalia in the male versus the internal (within the pelvic cavity) location of the major reproductive organs in the female.

Here's the main difference—males have major external genitalia but most female reproductive organs are inside the pelvic cavity.

Male reproductive system

The male reproductive system consists of the organs that produce, conduct and introduce mature sperm into the female reproductive tract, where fertilisation occurs. (See *Structures of the male reproductive system*.)

Extra work

In addition to forming male sex cells (spermatogenesis), the male reproductive system plays a role in the secretion of male sex hormones.

Zoom in

Structures of the male reproductive system

The male reproductive system consists of the penis, the scrotum and its contents, the prostate gland and the inguinal structures.

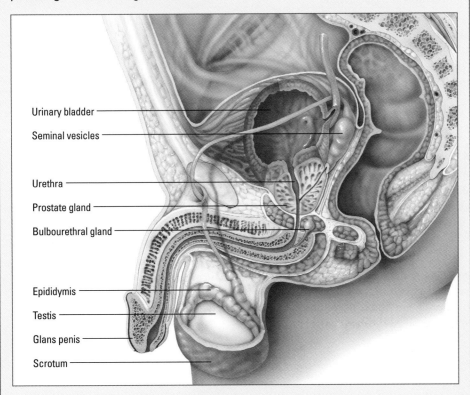

- Urinary bladder
- Seminal vesicles
- Urethra
- Prostate gland
- Bulbourethral gland
- Epididymis
- Testis
- Glans penis
- Scrotum

Penis

The organ of copulation, the *penis* deposits sperm in the female reproductive tract and acts as the terminal duct for the urinary tract. It consists of an attached root, a free shaft, and an enlarged tip or glans penis.

From the inside . . .

Internally, the cylinder-shaped penile shaft consists of three columns of *erectile tissue* bound together by *heavy fibrous tissue*. Two *corpora cavernosa* form the major part of the penis. On the underside, the *corpus spongiosum* encases the urethra. The enlarged proximal end of the urethra forms the bulb of the penis.

The *glans penis*, at the distal end of the shaft, is a cone-shaped structure formed from the corpus spongiosum. Its lateral margin forms a ridge of tissue known as the *corona*. The glans penis is highly sensitive to sexual stimulation.

. . . out

Thin, loose skin covers the penile shaft. The *urethral meatus* opens through the glans to allow urination and ejaculation.

In a different vein

The penis receives blood through the *internal pudendal artery*. Blood then flows into the corpora cavernosa through the penile artery resulting in erection. Venous blood returns through the *internal iliac vein* to the *vena cava*.

Scrotum

Located posterior to the penis and anterior to the anus, the scrotum is an extra-abdominal pouch that consists of a thin layer of skin overlying a tighter, muscle-like layer. This muscle-like layer overlies the tunica vaginalis, a serous membrane that covers the internal scrotal cavity.

Canals and rings

Internally, a *septum* divides the scrotum into two sacs, which each contain a *testis*, an *epididymis*, and a *spermatic cord*. The spermatic cord is a connective tissue sheath that encases autonomic nerve fibres, blood vessels, lymph vessels and the *vas deferens* (also called the *ductus deferens*).

The spermatic cord travels from the testis through the inguinal canal through which the testes descend in the foetus, exiting the scrotum through the external inguinal ring and entering the abdominal cavity through the internal inguinal ring. The inguinal canal lies between the two rings.

Loads of nodes

Lymph nodes from the penis, scrotal surface and anus drain into the *inguinal* lymph nodes. Lymph nodes from the testes drain into the lateral aortic and pre-aortic lymph nodes in the abdomen.

An important function of the scrotum is to keep the testes cooler than the rest of the body.

Testes

The testes are enveloped in two layers of connective tissue: the *tunica vaginalis* (outer layer) and the *tunica albuginea* (inner layer). Extensions of the tunica albuginea separate each testis into lobules. Each lobule contains one to four *seminiferous tubules*, small tubes in which spermatogenesis takes place.

Climate control

Spermatozoa development requires a temperature lower than that of the rest of the body. The *dartos muscle*, a smooth muscle in the superficial fascia, causes scrotal skin to wrinkle, which helps to regulate temperature. The *cremaster muscles*, rising from the internal oblique muscle, help to govern temperature by elevating the testes.

Duct system

The male reproductive *duct system*, consisting of the epididymis and vas deferens, conveys sperm from the testes to the ejaculatory ducts near the bladder.

Storage area

The *epididymis* is a coiled tube located superior to and along the posterior border of the testis. During ejaculation, smooth muscle in the epididymis contracts, ejecting spermatozoa into the vas deferens.

Descending and merging

The *vas deferens* leads from the testes to the abdominal cavity, where it extends upward through the *inguinal canal*, arches over the urethra and descends behind the bladder. Its enlarged portion, called the *ampulla*, merges with the duct of the seminal vesicle to form the short ejaculatory duct. Inside the prostate gland the ejaculatory ducts join with the urethra.

Tubular exit

A small tube leading from the floor of the bladder to the exterior, the *urethra* consists of three parts:
- *prostatic urethra*, which is closest to the bladder and is surrounded by the prostate
- *membranous urethra*, which passes through the urogenital diaphragm
- *spongy urethra*, which makes up about 75% of the entire urethra.

Accessory reproductive glands

The accessory reproductive glands, which produce most of the semen, include the *seminal vesicles, bulbourethral glands (Cowper's glands)* and the *prostate gland.* The seminal vesicles are paired sacs at the base of the bladder. The bulbourethral glands, also paired, are located inferior to the prostate.

Size of a walnut

The walnut-sized prostate gland lies under the bladder and surrounds the urethra. It consists of three lobes: the left and right lateral lobes and the median lobe.

Improving the odds

The prostate continuously secretes alkaline prostatic fluid that adds volume to the semen during sexual activity. The fluid functions to enhance sperm motility and may increase the chances for conception by neutralising the acidic environment of the male urethra and that of the vagina.

Slightly alkaline

Semen is a viscous secretion with a slightly alkaline pH (7.8 to 8); it consists of spermatozoa and accessory gland secretions. The seminal vesicles produce roughly 60% of the fluid portion of the semen, while the prostate gland produces about 30%. A viscid fluid secreted by the bulbourethral glands also becomes part of the semen.

Prostatic fluid enhances sperm motility and may increase the chances for conception by neutralising the acidity of the male urethra and the female reproductive tract.

Spermatogenesis

Sperm formation, or *spermatogenesis*, begins when a male reaches *puberty* and normally continues throughout life.

Divide and conquer

Spermatogenesis occurs in four stages:

In the first stage, the primary germinal epithelial cells, called *spermatogonia*, grow and develop into primary *spermatocytes*. Both spermatogonia and primary spermatocytes contain 46 chromosomes, consisting of 44 *autosomes* and the two sex chromosomes, X and Y.

Next, primary spermatocytes divide to form secondary spermatocytes. No new chromosomes are formed in this stage; the pairs only divide. Each secondary spermatocyte contains one-half the number of autosomes, 22. One secondary spermatocyte contains an X chromosome; the other, a Y chromosome.

In the third stage, each secondary spermatocyte divides again to form *spermatids* (also called *spermatoblasts*).

Finally, the spermatids undergo a series of structural changes that transform them into mature *spermatozoa*, or sperm. Each spermatozoa has a head, neck, body and tail. The head contains the *nucleus*; the tail a large amount of *adenosine triphosphate*, which provides energy for sperm *motility*.

Queuing up

New sperm pass from the seminiferous tubules through the *rete testes* into the epididymis, where they mature. Only a small number of sperm can be stored in the epididymis. Most of them move into the vas deferens, where they're stored until sexual stimulation triggers emission.

Keeps for weeks

Sperm cells retain their potency in storage for many weeks. After ejaculation, sperm can survive for up to 4 days in the female reproductive tract.

Memory jogger

To remember the meaning of *spermatogenesis*, keep in mind that *genesis* means 'beginning' or 'new'. Therefore, *spermatogenesis* means *beginning of new sperm*.

I'm now mature and ready to fertilise.

Male hormonal control and sexual development

Androgens (male sex hormones) are produced mainly in the testes although a lesser amount are produced in the adrenal glands. They are responsible for the development of male sex organs and secondary sex characteristics. The main androgen is testosterone. It is regulated by luteinising hormone (LH) and follicle-stimulating hormone (FSH) from the anterior pituitary gland.

Number and motility affect fertility. A low sperm count (less than 20 million per millilitre of ejaculated semen) may result in infertility.

The captain of the team

Interstitial (Leydig) cells, located in the testes between the seminiferous tubules, secrete *testosterone*, the most significant male sex hormone. Testosterone is responsible for the development and maintenance of male sex organs and secondary sex characteristics, such as facial hair and vocal cord thickness. Testosterone is also required for spermatogenesis.

Other hormones, produced by *Sertoli cells* located in the seminiferous tubules, assist in development and nurture of sperm.

Directing the game

Testosterone secretion begins approximately 2 months after conception, when the release of chorionic gonadotropins from the placenta stimulates interstitial (Leydig) cells in the male foetus. The presence of testosterone directly affects sexual differentiation in the foetus. With testosterone, foetal genitalia develop into a penis, scrotum and testes; without testosterone, genitalia develop into a clitoris, vagina and other female organs.

During the last 2 months of gestation, testosterone normally causes the testes to descend into the scrotum. If the testes don't descend after birth, exogenous testosterone may correct the problem.

No gonadotropins yet.

Other key players

Other hormones also affect male sexuality. Two of these, *LH* (also called *interstitial cell-stimulating hormone*) and *FSH*, directly affect secretion of testosterone.

Time to grow

During early childhood, gonadotropins aren't secreted and there is little circulating testosterone. Secretion of gonadotropins from the pituitary gland usually occurs between ages 11 and 14, and marks the onset of puberty. These pituitary gonadotropins stimulate testis functioning as well as testosterone secretion.

From boy to man

During puberty, the penis and testes enlarge and the male reaches full adult sexual and reproductive capability. Puberty also marks the development of

As time goes by . . .

Male reproductive changes with aging

Physiological changes in older men include reduced testosterone production which, in turn, may cause decreased libido. A reduced testosterone level also causes the testes to atrophy and softens and decreases sperm production by up to 70% between ages 60 and 80.

Normally, the prostate gland enlarges with age and its secretions diminish. Seminal fluid also decreases in volume and becomes less viscous.

Sexual changes

During intercourse, older men experience slower and weaker physiological reactions. However, these changes don't necessarily lessen sexual satisfaction.

---

male secondary sexual characteristics: distinct body hair distribution, skin changes (such as increased secretion by sweat and sebaceous glands), deepening of the voice (from laryngeal enlargement), increased musculoskeletal development and other intracellular and extracellular changes.

Reaching the plateau

After a male achieves full physical maturity, usually by age 20, sexual and reproductive function remain fairly consistent throughout life although sperm production falls significantly in those over age 60. (See *Male reproductive changes with aging*.)

Female reproductive system

Unlike the male reproductive system, the female system is largely internal, housed within the pelvic cavity. It's composed of the external genitalia, vagina, cervix, uterus, uterine tubes and ovaries.

The mammary glands are accessory organs of the female reproductive system that are specialised to secrete milk following pregnancy.

External genitalia

The vulva consists of the external female genitalia, those visible on inspection, including the *mons pubis, labia majora, labia minora, clitoris, vaginal opening* and adjacent structures. (See *Female external genitalia*, page 258.)

At the bottom

The *mons pubis* is a rounded cushion of fatty and connective tissue covered by skin and coarse hair, which lies over the symphysis pubis (the joint formed by the union of the pubic bones anteriorly).

Zoom in

Female external genitalia

The female external genitalia include the vaginal opening, mons pubis, clitoris, labia majora, labia minora and associated glands.

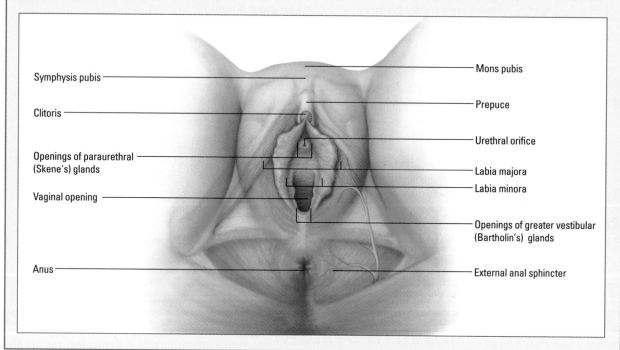

Symphysis pubis

Clitoris

Openings of paraurethral (Skene's) glands

Vaginal opening

Anus

Mons pubis

Prepuce

Urethral orifice

Labia majora

Labia minora

Openings of greater vestibular (Bartholin's) glands

External anal sphincter

Major league

The *labia majora* are two raised folds of skin that border the vulva on either side, extending from the mons pubis to the perineum. They are *homologous* (are similar in structure) to the scrotum and are covered by pubic hair. They contain an abundance of adipose and connective tissue, and possess numerous sebaceous (oil) and apocrine sudoriferous (sweat) glands.

Minor . . .

The *labia minora* are two moist folds of mucosal tissue that lie medial to the labia majora. The labia minora join anteriorly to form the *prepuce*, a hood-like covering over the clitoris, and posteriorly to form the *fourchette*, a thin tissue fold along the edge of the perineum.

. . . in name only

The labia minora contain numerous sebaceous glands, which secrete a lubricant that also acts as a bactericide. Like the labia majora, they are rich in blood vessels and nerve endings, which makes them highly responsive to stimulation. They swell in response to sexual stimulation, a reaction that triggers other changes that prepare the genitalia for coitus.

Small but sensitive

The *clitoris* is the small, protuberant organ just beneath the arch of the mons pubis. It contains erectile tissue, venous cavernous spaces and specialised sensory corpuscles, which are stimulated during sexual activity.

Mucho mucus

The *vestibule* is an oval area bounded anteriorly by the clitoris, laterally by the labia minora and posteriorly by the fourchette. The mucus-producing *Skene's glands* are found on both sides of the urethral opening. Openings of the two mucus-producing *Bartholin's glands* are located laterally and posteriorly on either side of the inner vaginal orifice.

The *urethral meatus* is the slitlike opening below the clitoris through which urine leaves the body. In the centre of the vestibule is the *vaginal orifice* which is bordered by the *hymen*, a tissue membrane.

The labia are highly vascular and have many nerve endings—making them sensitive to pain, pressure, touch, sexual stimulation, and temperature extremes.

Not too simple

Located between the lower vagina and the anal canal, the *perineum* is a complex structure of muscles, blood vessels, fasciae, nerves and lymphatics.

Vagina

The female internal genitalia, beginning with the vagina, are specialised organs; their main function lies in various aspects of the overall reproductive process. (See *Structures of the female reproductive system*, page 260.)

Three layers . . .

The *vagina*, a highly elastic muscular tube, is located between the urethra and the rectum. The vaginal wall has three tissue layers: epithelial tissue, loose connective tissue and muscle tissue. The *uterine cervix* connects the uterus to the vaginal vault. A recess, called the *fornix* surrounds the vaginal attachment to the cervix.

. . . three functions

The vagina has three main functions:
• accommodating the penis during coitus
• channelling blood discharged from the uterus during menstruation
• serving as the birth canal during childbirth.

Zoom in

Structures of the female reproductive system

The female reproductive system includes the vagina, cervix, uterus, uterine tubes, ovaries and other structures.

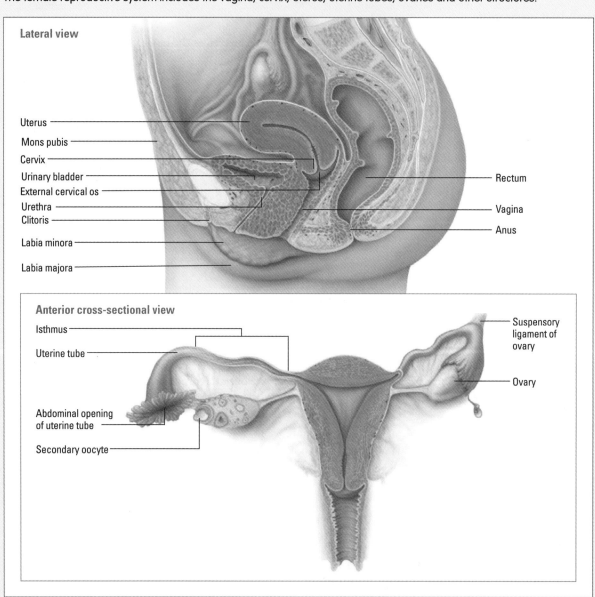

Lateral view

Uterus
Mons pubis
Cervix
Urinary bladder
External cervical os
Urethra
Clitoris
Labia minora
Labia majora

Rectum
Vagina
Anus

Anterior cross-sectional view

Isthmus
Uterine tube
Abdominal opening of uterine tube
Secondary oocyte

Suspensory ligament of ovary
Ovary

Supplied separately

The upper, middle and lower vaginal sections have separate blood supplies. Branches of the uterine arteries supply blood to the upper vagina, the *inferior vesical arteries* supply blood to the middle vagina, and the *haemorrhoidal* and *internal pudendal arteries* feed into the lower vagina.

Blood returns through a vast venous *plexus* to the haemorrhoidal, pudendal and uterine veins and then to the *hypogastric* veins. This plexus merges with the *vertebral venous plexus*.

Cervix

The cervix projects into the upper portion of the vagina. The lower cervical opening is the *external os*; the upper opening is the *internal os*.

Permanent alterations

Childbirth permanently alters the cervix. In a female who hasn't delivered a child, the external os is a round opening about 3 mm in diameter; after the first childbirth, it becomes a small transverse slit with irregular edges.

Uterus

The *uterus* is a small, firm, pear-shaped, muscular organ that's situated between the bladder and rectum. It typically lies over the urinary bladder at nearly a 90-degree angle to the vagina. The mucous membrane lining of the uterus is called the *endometrium*, and the muscular layer of the uterus is called the *myometrium*.

Fundamental fundus

During pregnancy, the elastic, upper portion of the uterus, called the *fundus*, accommodates most of the growing foetus until term. The uterine neck joins the fundus to the cervix, the part of the uterus that extends into the vagina. The fundus and neck make up the *corpus*, the main uterine body.

Uterine tubes

Two *uterine tubes* attach to the uterus at the upper angles of the fundus. These narrow cylinders of muscle fibres are the site of fertilisation.

Riding the wave

The curved portion of the uterine tube, called the *ampulla*, ends in the funnel-shaped *infundibulum*. Finger-like projections in the infundibulum, called *fimbriae*, move in waves that sweep the secondary oocyte from the ovary into the uterine tube.

> Fingerlike projections called fimbriae move in waves, sweeping the secondary oocyte from the ovary to the uterine tube.

Ovaries

The *ovaries* are located on either side of the uterus. The size, shape and position of the ovaries vary with age. Round, smooth, at birth, they grow larger, flatten, by puberty. During the childbearing years, they take on an almond shape and a rough, pitted surface; after menopause, they significantly reduce in size.

2,000,000 oocytes!

The ovaries' main function is to produce secondary oocytes ready for fertilisation. At birth, each ovary contains approximately 2 million follicles each containing a primary oocyte. The number of follicles reduces to around 250,000 by puberty and 400 of these mature into Graafian follicles, one of which will rupture, each month, to release a secondary oocyte under the influence of pituitary gonadotropins. If the oocyte isn't fertilised by sperm within approximately 1 day of ovulation, it will die. If it's fertilised, it will travel down a uterine tube to the uterus.

The ovaries also produce oestrogen, progesterone and a small amount of androgens.

Males and females have mammary glands—but they typically function only in the female.

Mammary glands

The mammary glands, which are located in the breasts, are specialised accessory glands that secrete milk. Although present in both sexes, they typically function only in the female.

Lobes, ducts and drainage

Each mammary gland contains 15 to 25 lobes separated by fibrous connective tissue and fat. Within the lobes are clustered acini—tiny, saclike duct terminals that secrete milk during lactation.

The ducts draining the lobules converge to form excretory (*lactiferous*) ducts and sinuses (*ampullae*), which store milk during lactation. These ducts drain onto the nipple surface through 15 to 20 openings. (See *The female breast.*)

Hormonal function and the menstrual cycle

Like the male body, the female body changes with age in response to hormonal control. (See *Events in the female reproductive cycle*, page 264.) When a female reaches the age of menstruation, the hypothalamus, ovaries and pituitary gland secrete hormones—*oestrogen, progesterone, FSH, and LH*—that affect the build-up and shedding of the endometrium during the menstrual cycle.

Zoom in

The female breast

The breasts are located on either side of the anterior chest wall over the greater pectoral and the anterior serratus muscles. Within the areola, the pigmented area in the centre of the breast, lies the nipple. Erectile tissue in the nipple responds to cold, friction and sexual stimulation.

Support and separate

Each breast is composed of glandular, fibrous and adipose tissue. Glandular tissue contains 15 to 20 lobes made up of clustered *acini*, tiny sac-like duct terminals that secrete milk. Fibrous suspensory ligaments *(Cooper's ligaments)* support the breasts; *adipose tissue* surrounds each breast.

Produce and drain

Acini draw the ingredients needed to produce milk from the blood in surrounding capillaries.

Sebaceous glands on the areolar surface *(Montgomery's glands)*, produce *sebum*, which lubricates the areolae and nipples during breast-feeding.

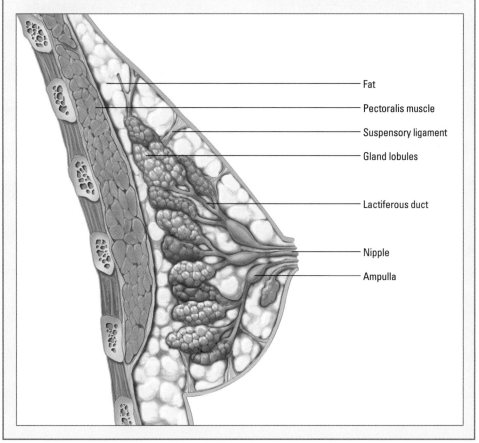

Fat

Pectoralis muscle

Suspensory ligament

Gland lobules

Lactiferous duct

Nipple

Ampulla

Now I get it!

Events in the female reproductive cycle

The female reproductive cycle averages 28 days. During this cycle, three major types of changes occur simultaneously: ovulatory, hormonal and endometrial (involving the lining [endometrium] of the uterus).

Ovulatory

- Ovulatory changes begin on the 1st day of the menstrual cycle.
- As the cycle begins, low oestrogen and progesterone levels in the bloodstream stimulate the hypothalamus to secrete gonadotropin-stimulating hormone (Gn-RH). In turn, Gn-RH stimulates the anterior pituitary gland to secrete follicle-stimulating hormone (FSH) and luteinising hormone (LH).
- Follicle development within the ovary (in the follicular phase) is spurred by increased levels of FSH and, to a lesser extent, LH.
- When the follicle matures, a spike in the LH level occurs, causing the follicle to rupture and release the ovum, thus initiating ovulation.
- After ovulation (in the luteal phase), the collapsed follicle forms the corpus luteum, which (if fertilisation doesn't occur) degenerates.

Hormonal

- During the follicular phase of the ovarian cycle, the increasing FSH and LH levels that stimulate follicle growth also stimulate increased secretion of oestrogen.
- Oestrogen secretion peaks just before ovulation. This peak sets in motion the spike in LH levels, which causes ovulation.

- After ovulation, oestrogen levels decline rapidly. In the luteal phase of the ovarian cycle, the corpus luteum is formed and beings to release progesterone and oestrogen.
- As the corpus luteum degenerates, levels of both of these ovarian hormones decline.

Endometrial

- The endometrium is receptive to implantation of an embryo for only a short time in the reproductive cycle. Thus, it's no accident that the endometrium is most receptive about 7 days after the initiation of ovulation—just in time to receive a fertilised ovum.
- In the first 5 days of the reproductive cycle, the endometrium sheds its functional layer, leaving the basal layer (the deepest layer) intact. Menstrual flow consists of this detached layer and accompanying blood from the detachment process.
- The endometrium begins regenerating its functional layer at about day 6 (the proliferative phase), spurred by rising oestrogen levels.
- After ovulation (about day 14), increased progesterone secretion stimulates conversion of the functional layer into a secretory mucosa (secretory phase), which is more receptive to implantation of the fertilised ovum.
- If implantation doesn't occur, the corpus luteum degenerates, progesterone levels drop and the endometrium again sheds its functional layer.

There are three major types of changes during the reproductive cycle: ovulatory, hormonal, and endometrial.

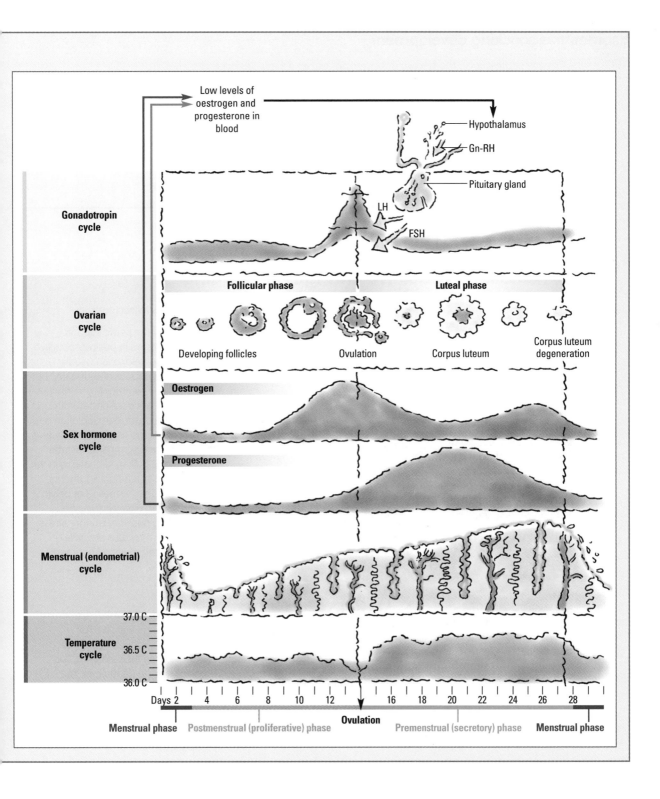

Low levels of oestrogen and progesterone in blood

Hypothalamus

Gn-RH

Pituitary gland

LH

FSH

Gonadotropin cycle

Ovarian cycle

Follicular phase

Luteal phase

Developing follicles

Ovulation

Corpus luteum

Corpus luteum degeneration

Sex hormone cycle

Oestrogen

Progesterone

Menstrual (endometrial) cycle

Temperature cycle

37.0 C
36.5 C
36.0 C

Days 2   4   6   8   10   12   16   18   20   22   24   26   28

Ovulation

Menstrual phase   Postmenstrual (proliferative) phase   Premenstrual (secretory) phase   Menstrual phase

A sometimes shocking development

During adolescence, the release of hormones causes a rapid increase in physical growth and spurs the development of secondary sex characteristics. This growth spurt begins at approximately age 11 and continues until early adolescence.

Menarche, the onset of menstruation, generally occurs after this growth spurt, usually between ages 11 and 14. Irregularity of the menstrual cycle is common during this time because of failure to ovulate. With menarche, the uterine body flexes on the cervix and the ovaries increase in size within the pelvic cavity.

As time goes by . . .

Female reproductive changes with aging

Declining oestrogen and progesterone levels cause numerous physical changes in aging women. Because women's breasts and internal and external reproductive structures are oestrogen-dependent, aging takes a more conspicuous toll on women than on men. As oestrogen levels decrease and menopause approaches, usually at about age 50, changes affect most parts of the female reproductive system. Significant emotional changes also take place during the transition from childbearing years to infertility.

Ovaries

Ovulation usually stops 1 to 2 years before menopause. As the ovaries reach the end of their productive cycle, they become unresponsive to gonadotropic stimulation. With aging, the ovaries atrophy and become thicker and smaller.

Vulva

The vulva also atrophies with age. Vulval tissue shrinks, exposing the sensitive area around the urethra and vagina to abrasions and irritations.

The vaginal orifice also constricts, tissues lose their elasticity and the epidermis thins from 20 layers to about 5. Other changes include pubic hair loss and flattening of the labia majora.

Vagina

Atrophy causes the vagina to shorten and the mucous lining to become thin, dry, less elastic and pale from decreased vascularity. In this state, the vaginal mucosa is highly susceptible to abrasion. In addition, the pH of vaginal secretions increases, making the vaginal environment more alkaline. The type of flora in it also changes, increasing older women's risk of vaginal infections.

Uterus

After menopause, the uterus shrinks rapidly to half its premenstrual weight. It continues to shrink until the organ reaches approximately one-fourth its premenstrual size. The cervix atrophies and no longer produces mucus for lubrication, and the endometrium and myometrium become thinner.

Pelvic support structures

Relaxation of the pelvic support commonly occurs in postreproductive women. Initial relaxation usually occurs during labour and delivery, but clinical effects commonly go unnoticed until the process accelerates with menopausal oestrogen depletion and loss of connective tissue elasticity and tone. Signs and symptoms include pressure and pulling sensations in the area above the inguinal ligaments, lower backaches, a feeling of pelvic heaviness and difficulty in rising from a sitting position. Urinary stress incontinence may also become a problem if urethrovesical ligaments weaken.

Breasts

In the breasts, glandular, supporting and fatty tissues atrophy. As suspensory ligaments (Cooper's ligaments) lose their elasticity, the breasts become pendulous. The nipples decrease in size and become flat, and the inframammary ridges (strands of supportive tissue that support the breast) become more pronounced.

A monthly thing

The menstrual cycle is a complex process that involves the reproductive and endocrine systems. The cycle averages 28 days in length.

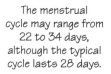

The menstrual cycle may range from 22 to 34 days, although the typical cycle lasts 28 days.

Supply exhausted

Cessation of menses usually occurs between ages 40 and 55. Although the pituitary gland still releases FSH and LH, the body has exhausted the supply of ovarian follicles that respond to these hormones, and menstruation no longer occurs.

A farewell to menses

A woman is considered to have reached menopause after menses are absent for 1 year. Before menopause, a woman experiences several transitional years (called the *climacteric years*), during which several physiological changes that lead to menopause occur. (See *Female reproductive changes with aging*.)

Quick quiz

1. Spermatogenesis is the:
    A. growth and development of sperm into primary spermatocytes.
    B. division of spermatocytes into secondary spermatocytes.
    C. passage of sperm into the epididymis.
    D. entire process of sperm formation.

Answer: D. Spermatogenesis refers to the entire process of sperm formation, from the development of primary spermatocytes to the formation of fully functional spermatozoa.

2. The primary function of the scrotum is to:
    A. provide storage for newly developed sperm.
    B. maintain a cool temperature for the testes.
    C. deposit sperm in the female reproductive tract.
    D. secrete prostatic fluid.

Answer: B. The scrotum maintains a cool temperature for the testes, which is necessary for spermatozoa formation.

3. The main function of the ovaries is to:
    A. secrete hormones that affect the build-up and shedding of the endometrium during the menstrual cycle.
    B. accommodate a growing foetus during pregnancy.
    C. produce ova.
    D. serve as the site of fertilisation.

Answer: C. The main function of the ovaries is to produce ova.

4. The corpus luteum forms and degenerates in which phase of the female reproductive cycle?
   A. luteal
   B. follicular
   C. proliferative
   D. secretory

Answer: A. The corpus luteum forms and degenerates in the luteal phase of the ovarian cycle.

5. The main hormones involved in the menstrual cycle are:
   A. LH, oestrogen, progesterone, testosterone and inhibin.
   B. oestrogen, FSH, LH and androgens.
   C. oestrogen, progesterone, LH and FSH and inhibin.
   D. gonadotropin-stimulating hormone, progesterone and testosterone.

Answer: C. The main hormones involved in the menstrual cycle are oestrogen, progesterone, LH, FSH and inhibin.

Scoring

☆☆☆ If you answered all five questions correctly, congratulations! You've hit a growth spurt in your knowledge.

☆☆ If you answered four questions correctly, good for you! You have a firm grasp of this complicated system.

☆ If you answered fewer than four questions correctly, keep trying! Like the reproductive system, your knowledge may just need some time to develop.

16 Reproduction and lactation

Just the facts

In this chapter, you'll learn:

♦ the process of fertilisation
♦ embryo and foetus development
♦ stages of labour
♦ the role of hormones in lactation.

Fertilisation

Creation of a new human being begins with *fertilisation*, the union of a *spermatozoon* and a *oocyte* (immature ovum) to form a single cell. After fertilisation occurs, dramatic changes begin inside a woman's body and in the oocyte. The cells of the fertilised ovum begin dividing as the ovum travels to the *uterine cavity*, where it implants in the uterine lining. (See *How fertilisation occurs*, page 270.)

Initiate

For fertilisation to take place, a spermatozoon must first reach the oocyte and trigger its development into a mature ovum. This process ensures the full complement of genetic material required for reproduction is available. Although a single ejaculation deposits several hundred-million spermatozoa, many are destroyed by acidic vaginal secretions. The only spermatozoa that survive are those that enter the *cervical canal*, where they're protected by *cervical mucus*.

> I'd better get moving! For fertilisation to take place, I have to reach the oocyte.

Timing is everything

The ability of spermatozoa to penetrate the cervical mucus depends on the phase of the menstrual cycle at the time of transit.

Early in the cycle, oestrogen and progesterone levels cause the mucus to thicken, making it more difficult for spermatozoa to pass through the cervix. During midcycle, however, when the mucus is relatively thin, spermatozoa can pass readily through the cervix. Later in the cycle, the cervical mucus thickens again, hindering spermatozoa passage.

Now I get it!

How fertilisation occurs

Fertilisation begins when a spermatozoon is activated upon contact with the oocyte (immature ovum). Here's what happens.

The spermatozoon head, which has a covering called the *acrosome,* approaches the oocyte.

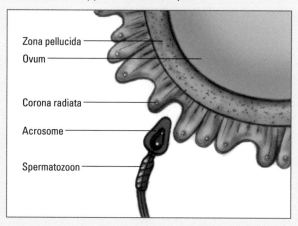

Zona pellucida
Ovum
Corona radiata
Acrosome
Spermatozoon

The acrosome develops small perforations through which it releases enzymes necessary for the sperm to penetrate the protective layers of the oocyte before fertilisation.

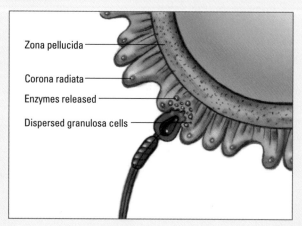

Zona pellucida
Corona radiata
Enzymes released
Dispersed granulosa cells

The spermatozoon then penetrates the zona pellucida (the inner membrane of the oocyte). This brings about development of the oocyte to a mature ovum. The penetrative action of the spermatozoon on the zona pellucida initiates chemical changes which prevent penetration by other spermatozoa. On penetration and completion of meiosis II, the oocyte becomes an ovum.

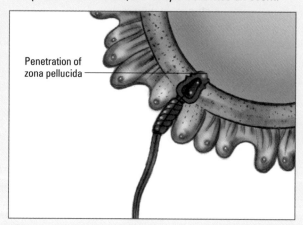

Penetration of zona pellucida

After the spermatozoon penetrates the oocyte, its nucleus is released into the oocyte, its tail degenerates and its head enlarges and fuses with the ovum's nucleus. This fusion provides the fertilised ovum, called a *zygote,* with 46 chromosomes.

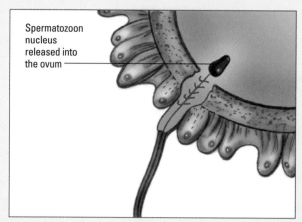

Spermatozoon nucleus released into the ovum

Help along the way

Spermatozoa travel through the female reproductive tract at a rate of several millimetres per hour by means of whiplike movements of the tail, known as *flagellar movements*.

After spermatozoa pass through the cervical mucus, however, the female reproductive system 'assists' them on their journey with rhythmic contractions of the uterus that help them penetrate the uterine tubes. Spermatozoa are typically viable (able to fertilise the ovum) for up to 2 days after ejaculation; however, they can survive in the female reproductive tract for up to 4 days.

A zygote is 'born'

Before a spermatozoon can penetrate the oocyte, it must disperse the *granulosa* cells (outer protective cells) and penetrate the *zona pellucida*—the thick, transparent layer surrounding the incompletely developed ovum. Enzymes in the *acrosome* (head cap) of the spermatozoon permit this penetration. After penetration, the oocyte undergoes development to become a mature ovum, and resulting chemical changes in the zona pellucida prevent penetration by other spermatozoa.

The head of the spermatozoon then fuses with the ovum nucleus, creating a cell nucleus with 46 chromosomes. The fertilised ovum is called a *zygote*.

Pregnancy

Pregnancy starts with fertilisation and ends with childbirth; on average, its duration is 38 to 40 weeks. During this period (called *gestation*), the zygote divides as it passes through the uterine tube and attaches to the uterine lining via implantation. A complex sequence of *pre-embryonic*, *embryonic* and *foetal* development transforms the zygote into a full-term foetus.

Making predictions

Because the uterus grows throughout pregnancy, uterine size serves as a rough estimate of gestation. The fertilisation date is rarely known and so a woman's expected delivery date is typically calculated from the beginning of their last menses. The tool used for calculating delivery dates is known as *Näegele's rule*.

Here's how it works: if you know the first day of the last menstrual cycle, simply count back 3 months from that date and then add 7 days. For example, let's say that the first day of the last menses was April 29. Count back 3 months, which gets you to January 29, and then add 7 days for an approximate due date of February 5.

Stages of foetal development

During pregnancy, the foetus undergoes three major stages of development:

- pre-embryonic period
- embryonic period
- foetal period.

Rite of passage

The pre-embryonic phase starts with fertilisation and lasts for 2 weeks. As the zygote passes through the uterine tube, it undergoes a series of *mitotic divisions*, or *cleavage*. (See *Pre-embryonic development*.)

Look, honey—it has your germ layers . . .

During the embryonic period (gestation weeks 3 through 8), the developing zygote starts to take on a human shape and is now called an *embryo*. Each germ layer—the *ectoderm*, *mesoderm* and *endoderm*—eventually forms specific tissues in the embryo. (See *Embryonic development*, page 274.)

. . . my organ systems . . .

The organ systems form during the embryonic period. During this time, the embryo is particularly vulnerable to injury by maternal drug use, certain maternal infections and other factors.

Memory jogger

To remember the three germ layers, keep in mind that *ecto* means *outside*, *meso* means *middle* and *endo* means *within*.

Now I get it!

Pre-embryonic development

The pre-embryonic phase lasts from conception until approximately the end of the second week of development.

Zygote formation . . .

As the ovum advances through the uterine tube towards the uterus, it undergoes mitotic division, forming daughter cells, initially called *blastomeres,* that each contain the same number of chromosomes as the parent cell. The first cell division ends about 30 hours after fertilisation; subsequent divisions occur rapidly.

The *zygote,* as it's now called, develops into a small mass of cells called a *morula,* which reaches the uterus at or around the third day after fertilisation. Fluid that amasses in the centre of the morula forms a central cavity.

. . . into blastocyst

The structure is now called a *blastocyst*. The blastocyst consists of a thin trophoblast layer, which includes the blastocyst cavity, and the inner cell mass. The trophoblast develops into foetal membranes and the placenta. The inner cell mass later forms the embryo *(late blastocyst).*

Getting attached: blastocyst and endometrium

During the next phase, the blastocyst stays within the zona pellucida, unattached to the uterus. The zona pellucida degenerates and, by the end of the first week after fertilisation, the blastocyst attaches to the endometrium. The part of the blastocyst adjacent to the inner cell mass is the first part to become attached.

The trophoblast, in contact with the endometrial lining, proliferates and invades the underlying endometrium by separating and dissolving endometrial cells.

Letting it all sink in

During the next week, the invading blastocyst sinks below the endometrium's surface. The penetration site seals, restoring the continuity of the endometrial surface.

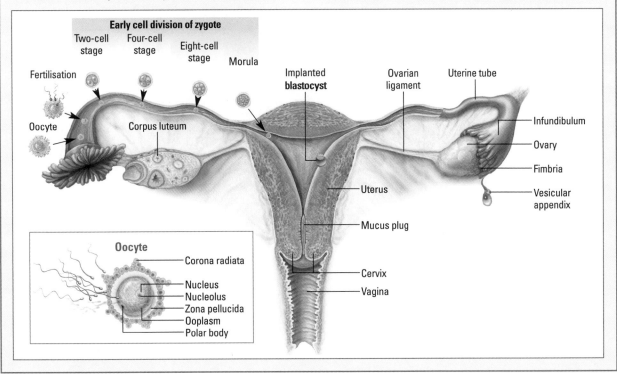

Now I get it!

Embryonic development

Each of the three germ layers—ectoderm, mesoderm and endoderm—forms specific tissues and organs in the developing embryo.

Ectoderm
The ectoderm, the outermost layer, develops into the:
- epidermis
- nervous system
- pituitary gland
- tooth enamel
- salivary glands
- optic lens
- lining of the lower portion of the anal canal
- hair.

Mesoderm
The mesoderm, the middle layer, develops into:
- connective and supporting tissue
- the blood and vascular system
- musculature
- teeth (except enamel)
- the mesothelial lining of the pericardial, pleural and peritoneal cavities
- the kidneys and ureters
- the gonads and associated ducts.

Endoderm
The endoderm, the innermost layer, becomes the epithelial lining of the:
- pharynx, larynx, trachea, bronchi and lungs
- auditory canal
- alimentary canal (except the oral cavity and anal canal)
- gall bladder, urinary bladder and liver
- thyroid, parathyroid, pancreas and thymus glands
- prostate, bulbourethral glands, vagina, vestibule, urethra and associated glands.

Mmmh . . . ectoderm is the only source of nervous and epithelial tissue, mesoderm is a unique source of muscle, and the endoderm forms the linings of the urinary and digestive tracts.

Chorionic villi

Embryonic disc

Ectoderm

Mesoderm

Endoderm

. . . and a very large head

During the *foetal* stage of development, which lasts from the 9th week until birth, the maturing foetus enlarges and grows heavier. (See *From embryo to foetus*.)

Now I get it!

From embryo to foetus

Significant growth and development take place within the first 3 months following fertilisation, as the embryo develops into a foetus that nearly resembles a full-term newborn.

Month 1

At the end of the first month, the embryo has a definite form. The head, the trunk and the tiny buds that will become the arms and legs are discernible. The cardiovascular system has begun to function, and the umbilical cord is visible in its most primitive form.

Month 2

During the second month, the embryo—called a foetus from the eighth week—grows to 2.5 cm in length and weighs 1 g. The head and facial features develop as the eyes, ears, nose, lips, tongue and tooth buds form.

The arms and legs also take shape. Although the gender of the foetus isn't yet discernible, all external genitalia are present. Cardiovascular function is complete, and the umbilical cord has a definite form. At the end of the second month, the foetus resembles a full-term newborn, except for size.

Month 3

During the third month, the foetus grows to 7.5 cm in length and weighs 28 g. Teeth and bones begin to appear, and the kidneys start to function. The foetus opens its mouth to swallow, grasps with its fully developed hands and prepares for

breathing by inhaling and exhaling amniotic fluid (although its lungs aren't functioning). At the end of the first *trimester* (the 3-month periods into which pregnancy is divided), the gender of the foetus is distinguishable.

Months 4 to 9

Over the remaining 6 months, foetal growth continues as internal and external structures develop at a rapid rate. In the third trimester, the foetus stores the fats and minerals it will need to live outside the womb. At birth, the average full-term foetus measures approximately 51 cm and weighs 3 to 3.5 kg.

1 month

2 months

3 months

9 months

Two unusual features appear during this stage:

👆 The head of the foetus is disproportionately large compared with its body. (This feature changes after birth as the infant grows.)

✌️ The foetus lacks subcutaneous fat. (Fat starts to accumulate shortly after birth.)

Structural changes in the ovaries and uterus

Pregnancy changes the usual development of the *corpus luteum* (the empty follicle that forms after ovulation and secretes progesterone and small amounts of oestrogen) and results in the development of the decidua, amniotic sac and fluid, yolk sac and placenta.

The foetus isn't the only thing changing during pregnancy— the reproductive system undergoes changes as well.

Corpus luteum

In the non-pregnant female, the normal functioning of the corpus luteum requires continued stimulation by luteinising hormone (LH). Progesterone produced by the corpus luteum suppresses LH release by the pituitary gland. If pregnancy occurs, the corpus luteum continues to produce progesterone until the placentas takes over. Otherwise, the copus luteum atrophies 3 days prior to the onset of menstruation.

Hormone soup

With age, the corpus luteum grows less responsive to LH. For this reason, the mature corpus luteum degenerates unless stimulated by progressively increasing amounts of LH.

Pregnancy stimulates the placental tissue to secrete large amounts of human chorionic gonadotropin (hCG) which is similar in structure and function to LH. The hCG prevents corpus luteum degeneration, stimulating the corpus luteum to produce the large amounts of oestrogen and progesterone needed to maintain the pregnancy during the first 3 months.

The hormone hCG can be detected as early as 7 days after fertilisation. It can confirm pregnancy before the first menstrual period is missed.

HCG tells all

As early as 7 days after fertilisation, hCG can be detected and can provide confirmation of pregnancy even before the first menstrual period is missed.

The hCG level gradually increases, peaks at about 10 weeks' gestation and then gradually declines.

Decidua

The *decidua* is the endometrial lining that undergoes the hormone-induced changes of pregnancy. Decidual cells secrete the following three substances:
- the hormone *prolactin*, which promotes lactation
- a peptide hormone, *relaxin*, which induces relaxation of the connective tissue of the symphysis pubis and pelvic ligaments and promotes cervical dilation
- a potent hormonelike fatty acid, *prostaglandin*, which mediates several physiological functions.

(See *Development of the decidua and foetal membranes*, page 278.)

Amniotic sac and fluid

The *amniotic sac*, enclosed within the *chorion* (an embryonic membrane that surrounds the embryo from the time of implantation), gradually enlarges and surrounds the embryo. As it grows, the amniotic sac expands into the chorionic cavity, eventually filling the cavity and fusing with the chorion by 8 weeks' gestation.

A warm, protective sea

The amniotic sac and amniotic fluid serve the foetus in two important ways—one during gestation and the other during delivery. During gestation, the fluid provides the foetus with a buoyant, temperature-controlled environment. Later, it serves as a fluid wedge that helps open the cervix during birth.

Thanks, mum—I'll take it from here

Early in pregnancy, amniotic fluid comes chiefly from three sources:

fluid filtering into the amniotic sac from maternal blood as it passes through the uterus

fluid filtering into the sac from foetal blood passing through the placenta

fluid diffusing into the amniotic sac from the foetal skin and respiratory tract.

Later in pregnancy, when the foetal kidneys begin to function, the foetus urinates into the amniotic fluid. Foetal urine then becomes the major component of amniotic fluid.

This is the life—warm, buoyant, and protected!

Daily gulps

Production of amniotic fluid from maternal and foetal sources compensates for amniotic fluid that's lost through the foetal GI tract. Normally, the foetus swallows up to several hundred millilitres of amniotic fluid each day. Some fluid is absorbed into the foetal circulation from the foetal GI tract; most

Development of the decidua and foetal membranes

Specialised tissues support, protect and nurture the embryo and foetus throughout its development. Among these tissues, the decidua and foetal membranes begin to develop shortly after conception.

Nesting place

During pregnancy, the endometrial lining is called the *decidua*. It provides a nesting place for the developing zygote and has some endocrine functions.

Based primarily on its position relative to the embryo (see below), the decidua may be known as the *decidua basalis,* which lies beneath the *chorionic vesicle;* the *decidua capsularis,* which stretches over the vesicle; or the *decidua parietalis,* which lines the rest of the endometrial cavity.

Network of blood vessels

The *chorion* is a membrane that forms the outer wall of the blastocyst. Vascular projections, called *chorionic villi,* arise from its periphery. As the *chorionic vesicle* enlarges, villi arising from the superficial portion of the chorion, called the *chorion laeve,* atrophy, leaving this surface smooth. Villi arising from the deeper part of the chorion, called the *chorion frondosum,* proliferate, projecting into the large blood vessels within the decidua basalis through which the maternal blood flows.

Blood vessels form within the villi as they grow and connect with blood vessels that form in the chorion, in the body stalk and within the body of the embryo. Blood begins to flow through this developing network of vessels as soon as the embryo's heart starts to beat.

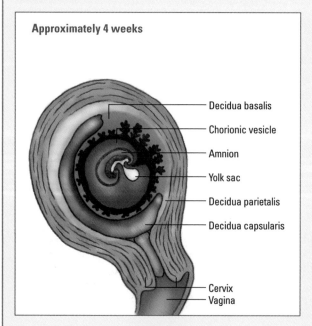

Approximately 4 weeks

- Decidua basalis
- Chorionic vesicle
- Amnion
- Yolk sac
- Decidua parietalis
- Decidua capsularis
- Cervix
- Vagina

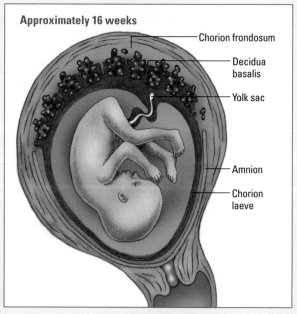

Approximately 16 weeks

- Chorion frondosum
- Decidua basalis
- Yolk sac
- Amnion
- Chorion laeve

is transferred from the foetal circulation to the maternal circulation and excreted in maternal urine.

Yolk sac

The *yolk sac* forms next to the endoderm; a portion of it is incorporated in the developing embryo and forms the GI tract. Another portion of the sac develops into primitive germ cells, which travel to the developing gonads and eventually form *oocytes* (precursors to ova) or *spermatocytes* (precursors to spermatozoa), after gender is determined at fertilisation.

No yolk—it's only temporary

During early embryonic development, the yolk sac also forms blood cells and vessels. Eventually, it atrophies and disintegrates.

Placenta

The flattened, disc-shaped *placenta*, using the umbilical cord as its conduit, provides nutrients to and removes wastes from the foetus from the third month of pregnancy until birth. The placenta is formed from the chorion, its chorionic villi and the adjacent decidua basalis.

The strongest link

The umbilical cord, which contains two arteries and one vein, links the foetus to the placenta. The umbilical arteries, which transport blood from the foetus to the placenta, take a spiral course on the cord, divide on the placental surface and branch off to the chorionic villi. (See *Picturing the placenta*.)

Zoom in

Picturing the placenta

At term, the *placenta* (the spongy structure within the uterus from which the foetus derives nourishment) is a flat, round or oval structure. It measures 15 to 20 cm in diameter and 2 to 3 cm in breadth at its thickest part. The maternal side is lobulated; the foetal side is smooth. After the birth, the placenta separates from the uterine wall and is expelled.

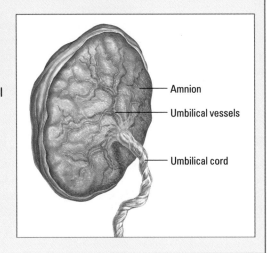

Amnion

Umbilical vessels

Umbilical cord

In a helpful vein

The placenta is a highly vascular organ. Large veins on its surface gather blood returning from the villi and join to form the single umbilical vein. The umbilical vein enters the cord, carrying oxygenated blood to the foetus.

Specialists on the job

The placenta contains two highly specialised circulatory systems:
- The *uteroplacental* circulation carries oxygenated arterial blood from the maternal circulation to the *intervillous spaces*—large spaces separating chorionic villi in the placenta. Blood enters the intervillous spaces from uterine arteries that penetrate the basal portion of the placenta; it leaves the intervillous spaces and flows back into the maternal circulation through veins in the basal portion of the placenta near the arteries.
- The *foetoplacental* circulation transports oxygen-depleted blood from the foetus to the chorionic villi through the umbilical arteries and returns oxygenated blood to the foetus through the umbilical vein.

Placenta takes charge

For the first 3 months of pregnancy, the corpus luteum is the main source of oestrogen and progesterone—steroid hormones required during pregnancy. By the end of the third month, however, the placenta produces most of the hormones; the corpus luteum which is no longer needed to maintain pregnancy, degenerates to form the corpus albicans.

Hormones on the rise

Levels of oestrogen and progesterone increase progressively throughout pregnancy. Oestrogen stimulates uterine development to provide a suitable environment for the foetus. It is also responsible for the maturation of various organs. Progesterone, synthesised by the placenta from maternal cholesterol, reduces uterine muscle irritability and prevents spontaneous abortion of the foetus.

Keep those acids coming

The placenta also produces *human placental lactogen* (HPL), which resembles growth hormone. HPL stimulates maternal protein and fat metabolism to ensure a sufficient supply of amino acids and fatty acids for the mother and foetus. HPL also stimulates breast growth in preparation for lactation. Throughout pregnancy, HPL levels rise progressively.

Labour and the postpartum period

Childbirth is achieved through labour—the process by which uterine contractions expel the foetus from the uterus. When labour begins, these contractions become strong and regular. Eventually, voluntary bearing-down efforts supplement the contractions, resulting in delivery. When that occurs, presentation of the foetus takes one of a variety of forms. (See *Comparing foetal presentations*, page 282.)

> Get ready to work! The contractions of labour are involuntary at first but are supplemented by voluntary efforts as birth approaches.

Onset of labour

The onset of labour results from several factors:
- The number of *oxytocin* (a pituitary hormone that stimulates uterine contractions) receptors on uterine muscle fibers increases progressively during pregnancy, peaking just before labour onset. This makes the uterus more sensitive to the effects of oxytocin.
- Stretching of the uterus over the course of the pregnancy initiates nerve impulses that stimulate oxytocin secretion from the posterior pituitary lobe.

Initiating start sequence

Near term, the foetal pituitary gland secretes more *adrenocorticotropic* hormone, which causes the foetal adrenal glands to secrete more *cortisol*. The cortisol diffuses into the maternal circulation through the placenta, heightens oxytocin and oestrogen secretion and reduces progesterone secretion. These changes intensify uterine muscle irritability and make the uterus even more sensitive to oxytocin stimulation.

Decline, diffuse and contract

Declining progesterone levels promote the formation of prostaglandins from arachidonic acid. These prostaglandins diffuse into the *uterine myometrium* inducing uterine contractions.

All systems go

As the cervix dilates, nerve impulses are transmitted to the central nervous system, causing an increase in oxytocin secretion from the posterior pituitary

Now I get it!

Comparing foetal presentations

Foetal presentations may be broadly classified as cephalic, breech, shoulder or compound.

Cephalic

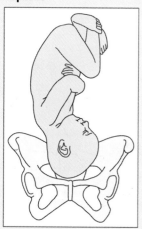

Breech

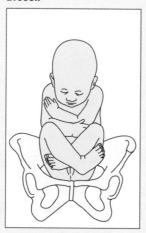

Shoulder

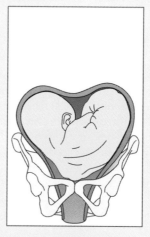

Compound

In the *cephalic*, or head-down presentation, the position of the foetus may be classified further by the presenting skull landmark, such as vertex (the topmost part of the head), brow, sinciput (the forward, upper part of the skull) or face.

In the *breech*, or head-up, presentation, the position of the foetus may be further classified as frank (hips flexed, knees straight), complete (knees and hips flexed), footling (knees and hips of one or both legs extended), kneeling (knees flexed and hips extended), or incomplete (one or both hips extended and one or both feet or knees lying below the breech).

Although a foetus may adopt one of several *shoulder* presentations, examination won't help a practitioner differentiate among them. Thus, all transverse positions are called *shoulder presentations.*

In a *compound* presentation, an extremity prolapses alongside the major presenting part so that two presenting parts appear at the pelvis at the same time.

gland. Acting as a positive feedback mechanism, increased oxytocin secretion stimulates more uterine contractions, which further dilate the cervix and lead the pituitary to secrete more oxytocin. Oxytocin secretions may also stimulate *prostaglandin* formation by the decidua.

Stages of labour

Childbirth can be divided into three stages. The duration of each stage varies according to the size of the uterus, the woman's age, and the number of previous pregnancies.

Stage 1—efface and dilate

The first stage of labour, in which the foetus begins its descent, is marked by cervical *effacement* (thinning) and *dilatation*. Before labour begins, the cervix isn't dilated; by the end of the first stage, it has dilated fully. The first stage of labour can last from 6 to 24 hours in primiparous women but is commonly significantly shorter for multiparous women. (See *Cervical effacement and dilatation*, page 284.)

Stage 2—no turning back

The second stage of labour begins with full cervical dilatation and ends with delivery of the foetus. During this stage, the amniotic sac ruptures as the uterine contractions increase in frequency and intensity. (The amniotic sac can also rupture before the onset of labour—during the first stage—and, although rare, sometimes ruptures after expulsion of the foetus.) As the flexed head of the foetus enters the pelvis, the mother's pelvic muscles force the head to rotate anteriorly and cause the back of the head to move under the symphysis pubis.

The curtain rises

As the uterus contracts, the flexed head of the foetus is forced deeper into the pelvis; resistance of the pelvic floor gradually forces the head to extend. As the head presses against the pelvic floor, vulvar tissues stretch and the anus dilates.

The star appears

The head of the foetus now rotates back to its former position after passing through the vulvovaginal orifice. Usually, head rotation is lateral (external) as the anterior shoulder rotates forwards to pass under the pubic arch. Delivery of the shoulders and the rest of the foetus follows. The second stage of labour averages about 45 minutes in primiparous women; it may be much shorter in multiparous women.

Curtain call

The third stage of labour starts immediately after childbirth and ends with placenta expulsion. After the neonate is delivered, the uterus continues to contract intermittently and grows smaller. The area of placental attachment

The first stage of labour can last from 6 to 24 hours in primiparous women (those with no previous history of a full-term pregnancy); the second stage lasts only about 45 minutes.

Both stages are typically much shorter for women who have delivered before.

Cervical effacement and dilation

Cervical effacement and dilation are significant aspects of the first stage of labour.

Thinning walls

Cervical effacement is the progressive shortening of the vaginal portion of the cervix and the thinning of its walls during labour as it's stretched by the foetus. Effacement is described as a percentage, ranging from 0% (noneffaced and thick) to 100% (fully effaced and paper-thin).

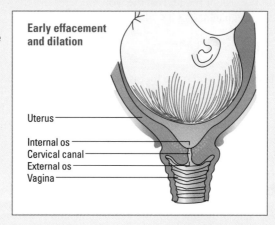

Early effacement and dilation

Uterus

Internal os
Cervical canal
External os
Vagina

Bigger exit

Cervical dilation refers to progressive enlargement of the *cervical os* to allow the foetus to pass from the uterus into the vagina. Dilation ranges from less than 1 cm to about 10 cm (full dilation).

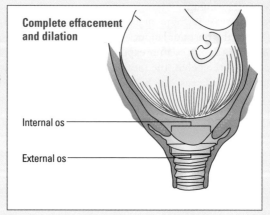

Complete effacement and dilation

Internal os

External os

also decreases. The placenta, which can't decrease in size, separates from the uterus, and blood seeps into the area of placental separation. The third stage of labour averages about 10 minutes in primiparous and multiparous women.

Postpartum period

After childbirth, the reproductive tract takes about 6 weeks to revert to its former condition during a process called *involution*. The uterus quickly grows smaller, with most of its involution taking place during the first 2 weeks after delivery.

Discharged with honours

Postpartum vaginal discharge (lochia) persists for several weeks after childbirth:
• *Lochia rubra*, a bloody discharge, occurs immediately after delivery and lasts for 1 to 4 days postpartum.
• *Lochia serosa*, a pinkish brown, serous discharge, occurs from 5 to 7 days postpartum.
• *Lochia alba*, a greyish white or colourless discharge, appears from 1 to 3 weeks postpartum.

Lactation

Lactation (milk synthesis and secretion by the breasts) is governed by interactions involving four hormones:
• oestrogen and progesterone, produced by the ovaries and placenta
• prolactin and oxytocin, produced by the pituitary gland under hypothalamic control.

Hormonal initiation of lactation

As gestation progresses, the production of oestrogen and progesterone by the placenta increases, causing glandular and ductal tissue in the breasts to proliferate. After breast stimulation by oestrogen and progesterone, prolactin causes milk secretion.

Pump priming

Oxytocin from the posterior pituitary gland causes contraction of specialised cells in the breast, producing a squeezing effect that forces milk down the ducts. Breast-feeding, in turn, stimulates prolactin secretion, resulting in a high prolactin level that induces changes in the menstrual cycle.

Opening the throttle

Progesterone and oestrogen levels fall after delivery. With oestrogen and progesterone no longer inhibiting prolactin's effects on milk production, the mammary glands start to secrete milk.

Pressing on the accelerator

Nipple stimulation during breast-feeding results in transmission of sensory impulses from the nipples to the hypothalamus. If the nipples aren't stimulated by breast-feeding, prolactin secretion declines after delivery.

Milk secretion continues as long as breast-feeding regularly stimulates the nipples. If breast-feeding stops, the stimulus for prolactin release is eliminated and milk production ceases.

Breast-feeding and the menstrual cycle

During the postpartum period, the presence of a high prolactin level inhibits FSH and LH release. If the mother does not breast-feed, prolactin output soon drops, ending inhibition of FSH and LH production by the pituitary. Subsequently, cyclic release of FSH and LH occurs. (See *Breast milk composition*.)

Regular breast-feeding provides the stimulus necessary for continued milk production.

A hold on ovulation

The menstrual cycle usually doesn't resume in breast-feeding women because prolactin inhibits the cyclic release of FSH and LH necessary for ovulation. This explains why breast-feeding initially protects against pregnancy.

The cycle comes full circle

Prolactin release in response to breast-feeding gradually declines, as does the inhibitory effect of prolactin on FSH and LH release. Consequently, ovulation and the menstrual cycle may resume. Pregnancy may occur after this, even with breast-feeding.

Breast milk composition

The composition of breast milk undergoes various changes during the process of lactation.

First tastes

Initial feedings provide a thin, serous fluid called colostrum. Unlike mature breast milk, which has a bluish tinge, colostrum is yellow. Colostrum contains high concentrations of protein, fat-soluble vitamins, minerals and immunoglobulins, which function as antibodies. Its laxative effect promotes early passage of meconium, the greenish black material that collects in the foetal intestines and forms the neonate's first stool. The breasts may contain colostrum for up to 96 hours after delivery.

From fore milk to hind milk

Breast milk composition continues to change over the course of a feeding. The fore milk—thin, watery milk secreted when a feeding begins—is low in calories but abounds in water-soluble vitamins. It accounts for about 60% of the total volume of a feeding. Next, hind milk is released. The hind milk, available 10 to 15 minutes after a breast-feeding session begins, has the highest concentration of calories; this helps to satisfy the neonate's hunger between feedings.

Quick quiz

1. Each of the three germ layers (ectoderm, mesoderm and endoderm) forms specific tissues and organs in the developing:
   A. zygote.
   B. ovum.
   C. embryo.
   D. foetus.

Answer: C. Each of the three germ layers (ectoderm, mesoderm and endoderm) forms specific tissues and organs in the developing embryo.

2. Which structure is responsible for protecting the foetus?
   A. Decidua
   B. Amniotic fluid
   C. Corpus luteum
   D. Yolk sac

Answer: B. Amniotic fluid, which provides a buoyant, temperature-controlled environment, protects the foetus.

3. Progressive enlargement of the cervical os during labour is called:
   A. dilatation.
   B. effacement.
   C. lactation.
   D. differentiation.

Answer: A. The progressive enlargement of the cervical os during labour is called *dilation*.

4. The initial breast milk that's yellow in colour is called:
   A. fore milk.
   B. hind milk.
   C. colostrum.
   D. prolactin.

Answer: C. The initial breast milk that's yellow in colour is called *colostrum*.

Scoring

 If you answered all four questions correctly, fabulous! You're first-rate with the physiology of fertilisation.

If you answered three questions correctly, excellent! You're looking good in the areas of labour and lactation.

If you answered fewer than three questions correctly, don't be alarmed! Cycling through the information in the chapter again should guarantee you won't reproduce those results.

Just the facts

In this chapter, you'll learn:

♦ the way genetic traits are transmitted

♦ the role of chromosomes and genes in heredity

♦ factors that determine trait predominance

♦ causes of genetic defects.

Your genes determine how you look, how your body functions and even whether you're prone to certain diseases.

A look at genetics

Genetics is the study of heredity—the passing of traits from biological parents to their children. People inherit not only physical traits, such as eye colour, but also biochemical and physiological traits, including the tendency to develop certain diseases. Genetic information is carried in *genes*, which comprise of groups of nucleotide bases arranged in a specific order within *deoxyribonucleic acid (DNA)*.

Family inheritance

Parents transmit inherited traits to their offspring through the germ cells *(gametes)*. There are two types of human gametes: eggs (ova) and sperm (spermatozoa).

Chromosomes

All cells (with the exception of mature erythrocytes) possess a nucleus which contains structures called *chromosomes*. Each chromosome contains a double strand of genetic material called *DNA*. This molecule is made up of thousands of nucleotide bases, which in a certain order constitute *genes*. These genes carry the code for proteins that influence each trait a person inherits, ranging from blood type to toe shape. Chromosomes exist in pairs except in the germ cells.

Counting chromosomes

A human ovum contains 23 chromosomes. A sperm also contains 23 chromosomes, each one a match to its partner in the ovum. When an ovum and a sperm unite, the corresponding chromosomes pair up resulting in a fertilised cell with 46 chromosomes (23 pairs) in its nucleus.

XX (or XY)

Of the 23 pairs of chromosomes in each living human cell, the two sex chromosomes of the 23rd pair determine a person's gender. The other 22 pairs are called *autosomes*.

In a female, both sex chromosomes are relatively large and each is designated by the letter X. In a male, one sex chromosome is an X chromosome and one is a smaller chromosome, designated by the letter Y.

Each gamete produced by a male contains either an X or a Y chromosome. Each gamete produced by a female contains an X chromosome. When a sperm with an X chromosome fertilises an ovum, the offspring is female (two X chromosomes). When a sperm with a Y chromosome fertilises an ovum, the offspring is male (one X and one Y chromosome).

Dividing the family assets

Ova and sperm are formed by a cell-division process called *meiosis*. In meiosis, each of the 23 pairs of chromosomes in a cell splits. The cell then divides, and each new cell (an ovum or sperm) receives one set of 23 chromosomes. (See Chapter 1 for more information on meiosis.)

I may look delicate, but my genotype says I'm strong!

Genes

Genes are segments of DNA, arranged in sequence on a chromosome. This sequence determines the properties of an organism.

Locus pocus

The location of a specific gene on a chromosome is called the *gene locus*. The locus of each gene is specific and doesn't vary from person to person. This allows each of the thousands of genes in an ovum to join the corresponding genes from a sperm when the chromosomes pair up at fertilisation.

How do I look?

The genetic information stored at a locus of a gene determines the genetic constitution—or *genotype*—of a person. The detectable, outward manifestation of a genotype is called the *phenotype*.

The genome at a glance

The human genome is made up of a double strand of deoxyribonucleic acid (DNA) molecules, one for each chromosome. Each strand comprises of thousands of chemical building blocks—*the nucleotide bases*—which, when arranged in specific sequence, constitute the 30,000 genes that are associated with these DNA molecules.

> If I were to read out loud the 3 billion bases found in just one genome, it would take over 9½ years.

The human genome

One complete set of chromosomes, containing all the genetic information for one person, is called a *genome*. (See *The genome at a glance*.)

Spilling the beans, on genes

For several years, scientists intensely studied the human genome to determine the entire sequence of each DNA molecule and the location and identity of all genes. The project was successfully completed in April 2003.

Catching the culprit

Genetic sequencing information allows practitioners to identify faulty genes that may cause disease. The benefits of this include future development of more specific diagnostic tests, the formation of new therapies and methods for avoiding conditions that trigger disease. It also allows clinicians and patients to consider and plan for a specific disease prognosis through genetic counselling.

Practitioners can now test patients for a gene error, present in 1 out of 500 people, that indicates an increased risk of developing colon cancer. Similarly, individuals with personal or family histories of breast or ovarian cancer can be tested for genetic predispositions to those diseases. Researchers are also seeking genes associated with dozens of other diseases, including chronic conditions such as asthma and diabetes.

> It only takes one bad gene to spoil a gene pool party.

Every

Each parent contributes one set of chromosomes (and therefore one set of genes) to their offspring. Therefore, each offspring has two genes for every locus (location on the chromosome) on the autosomal chromosomes.

Variation is the spice of life

Some characteristics, or traits, are determined by one gene that may have many variants. Variations of the same gene are called *alleles*. A person who has

Now I get it!

How genes express themselves

Genes account for inherited traits. *Gene expression* refers to a gene's effect on cell structure or function; however, the effects vary with the gene.

Dominant genes

Dominant genes can be transmitted and expressed by the offspring even if only one parent possesses the gene. Such genes mask the influence of the complementary gene on the homologous chromosome. Normal skin colour (as opposed to albinism) is a dominant trait. Astigmatism and familial hypercholesterolaemia are disease states governed by dominant genes.

Recessive genes

A recessive gene is expressed only when both parents transmit it to the offspring. Examples of recessive traits include cystic fibrosis, phenylketonuria and Tay Sachs disease.

Co-dominant genes

Co-dominant genes allow expression of both alleles and occur when the contributions of both alleles are visible in the phenotype. In ABO blood grouping, the *A* and *B* alleles are co-dominant in producing the AB blood group phenotype, in which both A- and B-type antigens are expressed.

Sex-linked genes

Sex-linked genes are carried on sex chromosomes. Almost all appear on the X chromosome and are recessive. In the male, sex-linked genes behave like dominant genes because no second X chromosome exists. Examples of X-linked recessive conditions include red-green colour blindness, haemophilia and Duchenne muscular dystrophy.

identical alleles on each chromosome is *homozygous* for that trait; if the alleles are different, they're said to be *heterozygous*. Other traits—called *polygenic traits*—require the interaction of more than one gene. Examples of polygenic traits include height, skin colour and weight. Polygenes allow a wide range of physical traits. For instance, height is regulated by several genes so that there will be a wide range of heights in a population.

Autosomal inheritance

On autosomal chromosomes, one allele may exert more influence in determining a specific trait. This is called the *dominant gene.* The less influential allele is called the *recessive gene.* Offspring express the trait of a dominant allele if both or only one chromosome in a pair carries it. For a recessive allele to be expressed, both chromosomes must carry recessive versions of the alleles. (See *How genes express themselves.*)

Sex-linked inheritance

The X and Y chromosomes are the sex chromosomes. The X chromosome is much larger than the Y. Therefore, males (who have XY chromosomes) have less genetic material than females (who have XX chromosomes), which means they have only one copy of most genes on the X chromosome. Inheritance of those genes is called *X-linked*, or *sex-linked*, *inheritance*.

Unequal X-change

A woman transmits one copy of each X-linked gene to each of her children, male or female. Because a man transmits an X chromosome only to his female children (male children receive a Y chromosome), he transmits X-linked genes only to his daughters, never his sons.

Multifactorial inheritance

Multifactorial inheritance reflects the interaction of at least two genes and the influence of environmental factors on expression.

Raising the bar

Height is a classic example of a multifactorial trait. In general, the height of an offspring peaks between the heights of the two parents. However, nutritional patterns, health care and other environmental factors also influence development of such traits as height. A better-nourished, healthier child of two short parents may be taller than either parent.

When it all comes together

Some diseases also have genetic predispositions for multifactorial inheritance; that is, the gene for a disease might be expressed only under certain environmental conditions.

Factors that may contribute to multifactorial inheritance include:
- use of drugs, alcohol or hormones by either parent
- maternal smoking
- maternal or paternal exposure to radiation
- maternal infection during pregnancy
- preexisting diseases in the mother
- nutritional factors
- general maternal or paternal health
- maternal-foetal blood incompatibility
- inadequate prenatal care.

> Good nutrition, proper health care and other environmental factors can influence the expression of a gene.

Genetic defects

Genetic defects are defects that result from changes to genes or chromosomes. They're categorised as either autosomal disorders, sex-linked disorders or multifactorial disorders. Some defects arise spontaneously, whereas others may be caused by environmental teratogens. *Teratogens* are environmental agents (such as infectious toxins, maternal diseases, drugs, chemicals and physical agents) that can cause structural or functional defects in a developing foetus. They may also cause spontaneous abortion, complications during labour and delivery, hidden defects in later development (such as cognitive or behavioural problems) and benign or cancerous tumours.

Permanent change in plans

A permanent change in genetic material is known as a *mutation*. Mutations can result from exposure to radiation, certain chemicals or viruses. They may also happen spontaneously and can occur anywhere in the genome.

Every cell has built-in defences against genetic damage. However, if a mutation isn't identified or repaired, it may produce a new trait that can be transmitted to offspring. Some mutations produce no effect, others change the expression of a trait and others can even change the way a cell functions. Some mutations cause serious or deadly defects, such as congenital anomalies and cancer.

Autosomal disorders

In autosomal disorders, an error occurs at a single gene site on the DNA strand. Single-gene disorders are inherited in clearly identifiable patterns that are the same as those seen in inheritance of normal traits. Because every person has 22 pairs of autosomes and only 1 pair of sex chromosomes, most hereditary disorders are caused by autosomal defects.

The assertive type

Autosomal dominant transmission involves transmission of an abnormal gene that's dominant. Autosomal dominant disorders usually affect male and female offspring equally. Children with one affected parent have a 50% chance of being affected.

Passive-aggressive behaviour

Autosomal recessive inheritance involves transmission of a recessive gene that's abnormal. Autosomal recessive disorders also usually affect male and female offspring equally. If both parents are affected, all their offspring will be affected. If both parents are unaffected but carry the defective gene, each child has a 25% chance of being affected. If only one parent is affected and the other isn't a carrier, none of the offspring will be affected, but all will carry the defective gene. If one parent is affected and the other is a carrier, each of their children will have a 50% chance of being affected. Because of this transmission pattern, autosomal recessive disorders may occur even when there's no family history of the disease.

Sex-linked disorders

Genetic disorders caused by genes located on the sex chromosomes are termed *sex-linked disorders*.

X calls the shots

Most sex-linked disorders are controlled by genes on the X chromosome, usually as recessive traits. Because males have only one X chromosome, a single X-linked recessive gene can cause disease to be exhibited in a male. Females receive two X chromosomes, so they may be homozygous for a disease allele (and exhibit the disease), homozygous for a normal allele (and neither have nor carry the disease) or heterozygous (carry, but not exhibit, the disease).

Most people who express X-linked recessive traits are males with unaffected parents. In rare cases, the father is affected and the mother is a carrier. All daughters of an affected male are carriers. An affected male never transmits the trait to his son. Unaffected male children of a female carrier don't transmit the disorder.

Evidence in history

With X-linked dominant inheritance, evidence of the inherited trait usually exists in the family history. A person with the abnormal trait must have one affected parent. If the father has an X-linked dominant disorder, all of his daughters and none of his sons will be affected. If a mother has an X-linked dominant disorder, each of her children has a 50% chance of being affected.

Multifactorial disorders

Most multifactorial disorders result from a number of genes and environmental influences acting together. In polygenic inheritance, each gene has a small additive effect and the combination of genetic errors is unpredictable. Multifactorial disorders can result from a less-than-optimum expression of many different genes, not from a specific error.

Mixing it up

Some multifactorial disorders are apparent at birth, such as cleft lip, cleft palate, congenital heart disease, anencephaly, clubfoot and myelomeningocele (a type of spina bifida). Others, such as type II diabetes mellitus, hypertension, hyperlipidaemia, most autoimmune diseases and many cancers, don't appear until later. Environmental factors most likely influence the development of multifactorial disorders during adulthood.

The majority of people who have X-linked recessive disorders are males. That's because males have only one X chromosome. Females have a second X chromosome that generally overpowers the 'diseased' X.

Now I get it!

Chromosomal disjunction and nondisjunction

These illustrations show disjunction and nondisjunction of an ovum. When disjunction proceeds normally, fertilisation with a normal sperm results in a zygote with the correct number of chromosomes. In nondisjunction, the duplicating chromosomes fail to separate; the result is one trisomic cell and one monosomic cell.

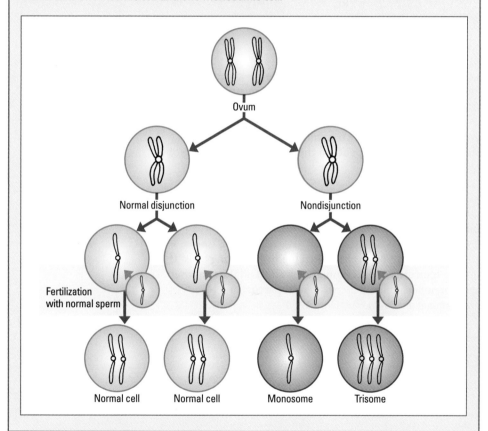

Chromosome defects

Aberrations in chromosome structure or number cause a class of disorders called *congenital anomalies*, or *birth defects*. These aberrations include the loss, addition or rearrangement of genetic material and are different from gene defects which arise from mutational events.

Most clinically significant chromosome aberrations arise during meiosis, an incredibly complex process that can go wrong in many ways. Potential contributing factors include maternal age, radiation and use of some therapeutic or recreational drugs.

Relocation, relocation

Translocation, the relocation of a segment of a chromosome to a nonhomologous chromosome, occurs when chromosomes split apart and rejoin in an abnormal arrangement. The cells still have a normal amount of genetic material, so often there are no visible abnormalities. However, the children of parents with translocated chromosomes may have serious genetic defects, such as monosomies or trisomies.

Unequal distribution

A *monosomy* is a condition in which the number of chromosomes present is one less than normal; an autosomal monosomy is incompatible with life. The presence of an extra chromosome is called a *trisomy* (the cause of such disorders as Down's syndrome). A mixture of both abnormal and normal cells results in *mosaicism* (two or more cell lines in the same person). The effects of mosaicism depend on the number and location of abnormal cells.

Breaking up is hard to do

During both meiosis and mitosis, chromosomes normally separate in a process called *disjunction*. Failure to separate, called *nondisjunction*, causes an unequal distribution of chromosomes between the two resulting cells. If nondisjunction occurs soon after fertilisation, it may affect all the resulting cells. The incidence of nondisjunction increases with parental age. (See *Chromosomal disjunction and nondisjunction*, page 295.)

Quick quiz

1. What's the total number of chromosomes in a fertilised cell?
    A. 12
    B. 23
    C. 46
    D. 52

Answer: C. There are 46 chromosomes (23 pairs) in the nucleus of a fertilised cell.

2.   According to genetic theory, if a child has cystic fibrosis (CF), an autosomal recessive disease, this must mean:
    A.   both parents transmit the gene for CF.
    B.   one parent transmits the gene for CF.
    C.   one grandparent has CF.
    D.   neither parent has the gene for CF.

Answer: A. Because the trait for CF is a recessive gene, it's only expressed when both parents transmit it to the offspring.

3.   The presence of an extra chromosome is called:
    A.   monosomy.
    B.   trisomy.
    C.   mosaicism.
    D.   nondisjunction.

Answer: B. Trisomy is the presence of an extra chromosome; it results from nondisjunction.

4.   Which definition applies to the term *mutation?*
    A.   an environmental agent responsible for a genetic defect
    B.   a permanent change in genetic material
    C.   interaction of at least two abnormal genes
    D.   expression of a recessive gene in an offspring

Answer: B. A mutation is a permanent change in genetic material that may result from exposure to radiation, certain chemicals or viruses. Mutations may also occur spontaneously.

5.   A child has brown eyes and brown hair. This description reveals the child's:
    A.   phenotype.
    B.   genotype.
    C.   genome.
    D.   autosomes.

Answer: A. Phenotype refers to the outward, detectable manifestation of a person's genetic makeup or genotype.

Scoring

★★★   If you answered all five questions correctly, excellent! When it comes to genetics, you're a gene-ius.

★★   If you answered four questions correctly, good job! Your understanding of genetics is definitely dominant.

★   If you answered fewer than four questions correctly, don't worry. Another look at the chapter may help to X-plain Y.

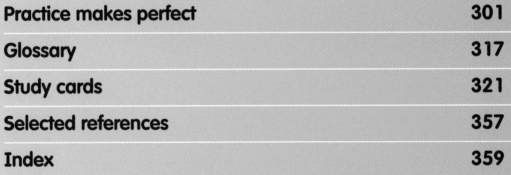

Appendices and index

1. The left atrium of the heart receives most of its blood from the:
   a. left subclavian artery.
   b. internal carotid artery.
   c. left coronary artery.
   d. right coronary artery.

2. The ventricles primarily receive blood during:
   a. inspiration.
   b. diastole.
   c. expiration.
   d. systole.

3. Which area of the heart is most likely to suffer an infarction after prolonged occlusion of the right coronary artery?
   a. Anterior
   b. Apical
   c. Inferior
   d. Lateral

4. Which valve prevents the backflow of blood from the left ventricle into the left atrium?
   a. Aortic
   b. Mitral
   c. Pulmonary
   d. Tricuspid

5. Stimulation of the sympathetic nervous system produces:
   a. bradycardia.
   b. tachycardia.
   c. hypotension.
   d. decreased myocardial contractility.

6. Which hormone can trigger an increase in blood pressure?
   a. Angiotensin I
   b. Angiotensin II
   c. Renin
   d. Parathyroid hormone

7. In response to a decrease in blood pressure, the hypothalamus secretes:
   a.   angiotensin.
   b.   antidiuretic hormone (ADH).
   c.   adrenaline (epinephrine).
   d.   renin.

8. An immature red blood cell (RBC) is called a:
   a.   B cell.
   b.   macrophage.
   c.   reticulocyte.
   d.   T cell.

9. The function of the thymus gland is to:
   a.   act as a reservoir for blood.
   b.   assist in the development of T cells.
   c.   protect the body against ingested pathogens.
   d.   remove bacteria and toxins from the circulatory system.

10. Reinflating collapsed alveoli improves oxygenation because:
    a.   alveoli require oxygen to remain viable.
    b.   reinflated alveoli decrease the demand for oxygen.
    c.   collapsed alveoli increase the demand for oxygen.
    d.   gas exchange occurs in the alveolar membrane.

11. A patient with an injury to the hypothalamus is most likely to experience which of
    these conditions?
    a.   Uncontrollable seizures
    b.   An infection
    c.   Tiredness
    d.   Sleeping difficulties

12. After a closed head injury, a patient develops memory loss and problems associated
    with the interpretation of sound. Which brain lobe was most likely injured?
    a.   Frontal
    b.   Occipital
    c.   Parietal
    d.   Temporal

13. Which connective tissue enables bones to move when skeletal muscles contract?
    a.   Tendon
    b.   Ligament
    c.   Adipose tissue
    d.   Nervous tissue

14. Osteoblast activity is necessary for:
    a.   bone formation.
    b.   oestrogen production.
    c.   haematopoiesis.
    d.   muscle development.

15. The closely packed, positively charged particles within an atom's nucleus are called:
    a. electrons.
    b. neutrons.
    c. protons.
    d. subatomic particles.

16. The hormones triiodothyronine (T_3) and thyroxine (T_4) are primarily involved in which body processes?
    a. Blood glucose level and glycogenesis
    b. Metabolic rate
    c. Growth of bones, muscles, and other organs
    d. Bone resorption, calcium absorption, and blood calcium levels

17. Which groups of hormones are released by the medulla of the adrenal gland?
    a. Adrenaline (epinephrine) and noradrenaline (norepinephrine)
    b. Glucocorticoids, mineralocorticoids, and androgens
    c. Thyroxine (T_4), triiodothyronine (T_3), and calcitonin
    d. Insulin, glucagon, and somatostatin

18. The anterior pituitary gland secretes which hormones? Select all that apply.
    a. Corticotropin
    b. Antidiuretic hormone (ADH)
    c. Thyroid-stimulating hormone (TSH)
    d. Prolactin

19. The thyroid gland is located in which area of the body?
    a. Upper abdomen
    b. Inferior aspect of the brain
    c. Upper portion of the kidney
    d. Lower neck, anterior to the trachea

20. The thyroid gland produces which hormones? Select all that apply.
    a. Triiodothyronine (T_3)
    b. Calcitonin
    c. Thyroxine (T_4)
    d. Thyroid-stimulating hormone (TSH)

21. One of the main functions of the skin is protection of inner body structures. One way the skin accomplishes this is to:
    a. repair surface wounds.
    b. impair the immune response.
    c. prevent the secretion of sebum.
    d. stop cell migration.

22. A patient complains of frequent burns while cooking because they do not feel hot temperatures. Which lobes of the patient's brain are most likely dysfunctional?
    a. Frontal
    b. Occipital
    c. Parietal
    d. Temporal

23. Emotional stress can cause the palms of the hand to sweat. Which gland is responsible for this occurrence?
    a. Pineal gland
    b. Eccrine gland
    c. Sebaceous gland
    d. Thyroid gland

24. During pregnancy, the foetus undergoes major stages. Place the following stages in order of their occurrence. Use all options.

a. Foetal period
b. Pre-embryonic period
c. Embryonic period

25. Which cells produce the pigment that contributes to hair colour?
    a. Keratinocytes
    b. Melanocytes
    c. Langerhans cells
    d. Merkel cells

26. The kidneys normally receive what percentage of the cardiac output?
    a. 10%
    b. 20%
    c. 30%
    d. 40%

27. Place the steps in order to trace the path of blood as it first becomes oxygenated and is delivered to the body. Use all of the options.

| a. Right ventricle |
| b. Left ventricle |
| c. Pulmonary arteries |
| d. Pulmonary veins |
| e. Left atrium |
| f. Aorta |

| |
| |
| |
| |
| |
| |

28. Using Näegele's rule, what date in October would a patient be due if they state the first day of their last menstrual cycle was January 1?

_____

29. Excess bicarbonate retention can result in:
    a. metabolic acidosis.
    b. respiratory acidosis.
    c. metabolic alkalosis.
    d. respiratory alkalosis.

30. Water-soluble vitamins include:
    a. Vitamin A.
    b. Vitamin C.
    c. Vitamin E.
    d. Vitamin K.

31. The prostate gland is a small gland located:
    a.   behind the bladder.
    b.   under the bladder and around the urethra.
    c.   along the posterior border of the testes.
    d.   inside the scrotum.

32. The amniotic sac fills the chorionic cavity by:
    a.   4 weeks' gestation.
    b.   6 weeks' gestation.
    c.   8 weeks' gestation.
    d.   10 weeks' gestation.

33. The function of the gallbladder is to:
    a.   produce enzymes that assist with digestion.
    b.   recycle bile salts.
    c.   remove bacteria from the blood.
    d.   store and concentrate bile.

34. Identify the area into which fluid would move during osmosis.

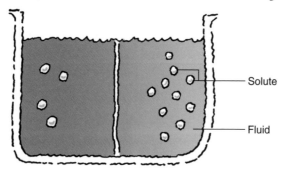

35. The average amount of urine produced daily is:
    a.   200 to 700 mL.
    b.   720 to 2,400 mL.
    c.   2,500 to 4,500 mL.
    d.   4,700 to 7,000 mL.

36. The continuous process whereby bone is created and destroyed is known as:
    a.   remodelling.
    b.   formation.
    c.   ossification.
    d.   classification.

37. Identify the carina in this illustration.

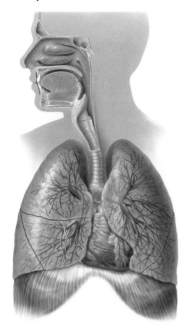

38. The connective tissue that covers and contours the spinal tissue and brain is known as the:
    a. endosteal dura.
    b. meningeal dura.
    c. pia mater.
    d. dura mater.

39. The cranial nerve (CN) responsible for the sensation of taste is the:
    a. olfactory nerve.
    b. trochlear nerve.
    c. facial nerve.
    d. hypoglossal nerve.

40. Which physiological changes affect nutrition in elderly patients? Select all that apply.
    a. Decreased renal function
    b. Decreased biting force
    c. Diminished enzyme activity and reduced gastric secretion
    d. Enhanced gag reflex

41. Transmission of sound vibrations to the middle ear is the function of which structure?
    a. Tympanic membrane
    b. Eustachian tube
    c. Auricle
    d. Vestibule

42. Which eye structure receives stimuli and sends them to the brain?
    a.   Lens
    b.   Sclera
    c.   Iris
    d.   Retina

43. Which type of blood cell participates in the body's defense and immune systems?
    a.   White blood cells (WBCs)
    b.   Red blood cells (RBCs)
    c.   Platelets
    d.   Thrombocytes

44. Moving a body part backward and forward is called:
    a.   eversion and inversion.
    b.   retraction and protraction.
    c.   flexion and extension.
    d.   pronation and supination.

45. A patient is experiencing problems with balance and fine and gross motor function. Identify the area of the patient's brain that is malfunctioning.

46. Acidity of the blood is determined by its pH. Which pH value indicates acidity?
    a.   7.24
    b.   7.35
    c.   7.44
    d.   7.54

47. Which hormones govern the lactation process?
    a.   Oestrogen and progesterone
    b.   Oestrogen and corticotropin
    c.   Follicle-stimulating hormone (FSH) and oestrogen
    d.   Growth hormone and progesterone

48. The movement of solutes from an area of higher concentration to one of lower concentration is called:
    a.   osmosis.
    b.   active transport.
    c.   diffusion.
    d.   endocytosis.

49. The hormone glucagon is produced in the pancreas by:
    a. Alpha cells.
    b. Beta cells.
    c. Delta cells.
    d. F cells.

50. Creatinine is a metabolic waste product excreted in urine and derived from:
    a. liver.
    b. muscle.
    c. bone.
    d. skin.

Answers

1. c. The left coronary artery, which splits into the anterior descending artery and the circumflex artery, is the primary source of blood for the left atrium. The left subclavian artery supplies blood to the arms, the internal carotid artery supplies blood to the head, and the right coronary artery supplies blood to the inferior wall of the heart.

2. b. At the beginning of diastole, the semilunar valves close to prevent backflow of blood into the ventricles, and the mitral and tricuspid valves open, allowing blood to flow into the ventricles from the atria. Breathing patterns (inspiration and expiration) are not related to blood flow.

3. c. The right coronary artery supplies blood to the right atrium, part of the left atrium, most of the right ventricle, and the inferior portion of the left ventricle. Therefore, prolonged occlusion could produce an infarction in the inferior area, although any areas that this cardiac artery supply may be affected. The right coronary artery does not supply blood to the anterior, lateral, or apical portions of the heart.

4. b. The mitral valve prevents backflow of blood from the left ventricle into the left atrium. The aortic valve prevents backflow from the aorta into the left ventricle. The pulmonary valve prevents backflow from the pulmonary artery into the right ventricle. The tricuspid valve prevents backflow from the right ventricle into the right atrium.

5. b. Stimulation of the sympathetic nervous system produces tachycardia and increased myocardial contractility. The other symptoms listed result from stimulation of the parasympathetic nervous system.

6. b. Angiotensin II exerts a powerful constricting effect on arterioles, causing blood pressure to rise. Angiotensin I is converted to angiotensin II by angiotensin-converting enzyme. Renin is an enzyme that leads to the formation of angiotensin I. The main function of parathyroid hormone is to help regulate serum calcium levels.

7. b. ADH acts on the renal tubules to promote water retention, which increases blood pressure. Although angiotensin, adrenaline (epinephrine), and renin also help increase blood pressure, they are not stored in the hypothalamus.

8. c. An immature RBC is called a *reticulocyte*. B cells, macrophages, and T cells are lymphocytes.

9. b. Bone marrow produces immature blood cells (stem cells). Those that become lymphocytes (T cells) migrate to the thymus to develop immunocompetence and tolerance in the foetus and then proceed to the lymph nodes and circulation. Lymphocytes are responsible for cell-mediated immunity. The spleen acts as a reservoir for blood cells. The tonsils shield against airborne and ingested pathogens. Lymph nodes remove bacteria and toxins from the bloodstream.

10. d. Gas exchange occurs in the alveolar membrane; therefore, collapsed alveoli decrease the surface area available for gas exchange, which decreases oxygenation of the blood. All alveoli, whether collapsed or not, receive oxygen and other nutrients from the bloodstream. Collapsed alveoli do not increase oxygen demand.

11. d. The hypothalamus helps regulate body temperature, appetite, water balance, pituitary secretions, emotions, and autonomic functions, including sleeping and waking cycles. Therefore, injury to the hypothalamus is most likely to be associated with sleeping difficulties. Neurotransmitter dysfunction is more likely to trigger seizures. A compromised immune system would make the patient tired and also place the patient at risk for infection.

12. d. The temporal lobe controls memory, acoustic interpretation, and language comprehension. The frontal lobe influences thinking, planning, and judgment. The occipital lobe regulates vision. The parietal lobe interprets sensations.

13. a. Tendons are bands of connective tissue that attach muscles to bone, enabling them to move when skeletal muscles contract. Ligaments are fibrous connective tissue that bind bones to other bones. Adipose tissue is loose connective tissue that insulates the body. Nervous tissue is not connective tissue.

14. a. Osteoblasts are bone-forming cells. Oestrogen contributes to the development of bone tissue through calcium reabsorption. Haematopoiesis (production of red blood cells) occurs in the bone marrow. Osteoblasts have no role in muscle development.

15. c. Protons are closely packed, positively charged particles within an atom's nucleus. Each element has a distinct number of protons. Electrons are negatively charged particles that orbit the nucleus in electron shells. Neutrons are uncharged, or neutral, particles in the atom's nucleus. Protons, neutrons, and electrons are all subatomic particles, but each has a different charge.

16. b. T_3 and T_4 are thyroid hormones that primarily affects metabolic rate. Bone resorption and increased calcium absorption are the principle effects of parathyroid hormone. Glucagon raises blood glucose levels and stimulates glycogenesis. The growth hormone somatotropin affects the growth of bones, muscles, and other organs.

17. a. The medulla of the adrenal gland releases adrenaline (epinephrine) and noradrenaline (norepinephrine). Glucocorticoids, mineralocorticoids, and androgens are released from the adrenal cortex. T_4, T_3, and calcitonin are secreted by the thyroid gland. The islet cells of the pancreas secrete insulin, glucagon, and somatostatin.

18. a, c, d. The anterior pituitary gland secretes corticotropin, TSH, and prolactin as well as FSH, growth hormone, and luteinising hormone. Inadequate secretion of corticotropin from the pituitary gland results in adrenal insufficiency. ADH is secreted by the posterior pituitary gland.

19. d. The thyroid gland resides in the lower neck, anterior to the trachea. The pancreas is in the upper abdomen. The pituitary gland is located in the inferior aspect of the brain. The adrenal glands are attached to the upper portion of the kidneys.

20. a, b, c. T_3, T_4, and calcitonin are all secreted by the thyroid gland. TSH is secreted by the pituitary gland.

21. a. The skin protects the body by intensifying normal cell replacement mechanisms to repair surface wounds. The epidermal layer of the skin contains Langerhans' cells that enhance the immune response by helping lymphocytes process antigens entering the skin. Sebum is secreted by the sebaceous glands; the skin does not prevent its secretion. Migration and shedding of cells help protect the skin.

22. c. The parietal lobes receive and appreciate sensations transmitted to it from the various sensory receptors in the body. This includes the ability to sense hot or cold objects. The frontal lobes regulate thinking, planning, and judgment. The occipital lobes interpret visual stimuli. The temporal lobes regulate memory.

23. b. The eccrine glands, also known as *sweat glands*, secrete fluid on the palms and soles of the feet in response to emotional stress. The pineal gland secretes melatonin which contributes to the regulation of the sleep cycle. The thyroid gland produces hormones that control and affect body function. The sebaceous glands produce sebum, which helps protect the skin's surface; they are located in all areas of the skin except on the hands and soles of the feet.

24.

b.	Pre-embryonic period
c.	Embryonic period
a.	Foetal period

The pre-embryonic phase starts with ovum fertilisation and lasts for 2 weeks. Weeks 3 through 8 encompass the embryonic period, during which time the developing zygote starts to take on a human shape and is called an *embryo*. The foetal stage of development lasts from week 9 until birth. During this period, the maturing foetus enlarges and grows heavier.

25. b. Hair colour is mainly influenced by melanin which is derived from melanocytes in the bulb matrix. Actual colour is determined by genes which direct the amount and type of melanin the melanocytes produce. Dark hair tends to possess more pure melanin while lighter shades have more melanin derivatives. Keratinocytes, which constitute 95% of the epidermal cell population, produce keratin which helps waterproof and protect the skin and underlying structures. Langerhans cells have an immune function and Merkel cells are involved in the sensation of touch.

26. b. The kidneys are highly vascular and receive about 25% of the blood pumped by the heart each minute.

27.

a.	Right ventricle
c.	Pulmonary arteries
d.	Pulmonary veins
e.	Left atrium
b.	Left ventricle
f.	Aorta

Unoxygenated blood travels from the right ventricle through the pulmonary valve into the pulmonary arteries. After passing into the lungs, it travels to the alveoli, where it exchanges carbon dioxide for oxygen. The oxygenated blood returns via the pulmonary veins to the left atrium. It then passes through the mitral valve and into the left ventricle, where it is pumped out to the body via the aorta.

28. Näegele's rule calculates a due date by counting back 3 months from the first day of the last menstrual cycle and then adding 7 days. If the first day of the patient's last period was January 1, they would be due on October 8.

29. c. Excess bicarbonate retention can result in metabolic alkalosis. Excess bicarbonate loss results in metabolic acidosis. Excess carbon dioxide retention results in respiratory acidosis. Excess carbon dioxide loss results in respiratory alkalosis.

30. b. Water-soluble vitamins include the B complex and C vitamins. Vitamins A, E, K, and D are fat-soluble vitamins.

31. b. The prostate gland lies under the bladder and surrounds the urethra. The vas deferens descends behind the bladder. The epididymis is located superior to and along the posterior border of the testes. The two sacs within the scrotum each contain a testis, an epididymis, and a spermatic cord.

32. c. The amniotic sac expands into the chorionic cavity, eventually filling the cavity and fusing with the chorion by 8 weeks' gestation.

33. d. The gallbladder stores and concentrates bile produced by the liver. The pancreas produces enzymes that assist with digestion. The gallbladder is not responsible for removing bacteria from blood. Bile salts are recycled by the liver.

34.

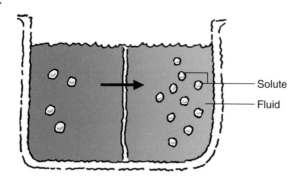

In osmosis, fluid moves passively from an area with more fluid (and fewer solutes) to one with less fluid (and more solutes).

35. b. Total daily urine output averages 720 to 2,400 ml per day; however, this amount varies with fluid intake and climate.

36. a. Remodelling is the continuous process whereby bone is created and destroyed. Formation refers to the development of bone. Ossification is bone hardening. Classification involves identifying bone according to its shape (long bone, short bone, or flat bone).

37.

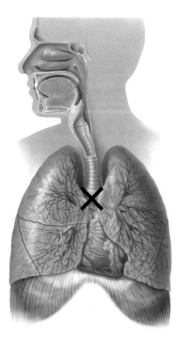

The carina is a ridge-shaped structure that is located at the level of the sixth or seventh thoracic vertebrae. It is the point where the trachea divides into the right and left main bronchi.

38. c. The pia mater is a continuous, delicate layer of connective tissue that covers and contours the spinal tissue and brain. The endosteal dura forms the periosteum of the skull and is continuous with the lining of the vertebral canal. The meningeal dura covers the brain, dipping between the brain tissue. The dura mater is leather-like tissue composed of the endosteal dura and meningeal dura.

39. c. The facial nerve (CN VII) controls taste and facial muscle movement. The olfactory nerve (CN I) is responsible for the sense of smell. The trochlear nerve (CN IV) controls extraocular eye movement. The hypoglossal nerve (CN XII) controls tongue movement.

40. a, b, c. Physiological changes that affect nutrition status in elderly patients include decreased renal function, decreased biting force, diminished enzyme activity, and reduced gastric secretion. Other changes include diminished intestinal activity, diminished sense of taste, diminished gag reflex, and decreased salivary flow.

41. a. The tympanic membrane—which consists of layers of skin, fibrous tissue, and a mucous membrane—transmits sound vibrations to the middle ear. The eustachian tube allows the pressure against inner and outer surfaces of the tympanic membrane to equalise, preventing rupture. The auricle is part of the external ear. The vestibule is the entrance to the inner ear.

42. d. The retina is the innermost coat of the eyeball; it receives visual stimuli and sends them to the brain. The lens refracts and focuses light onto the retina. The sclera helps maintain the size and form of the eyeball. The iris contains muscles and has an opening in the centre for the pupil, to allow light entry.

43. a. WBCs, which are classified as granulocytes or agranulocytes, participate in the body's defence and immune systems. RBCs transport oxygen and carbon dioxide to and from body tissues. Platelets, also known as *thrombocytes*, are involved in blood coagulation.

44. b. Retraction and protraction refer to backward and forward movement of a joint. Eversion and inversion involve moving a joint outward and inward. Flexion and extension involve increasing or decreasing a joint angle. Pronation and supination involve turning a body part downward or upward.

45.

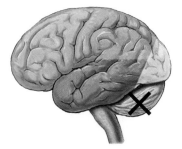

The cerebellum is the portion of the brain that controls balance and fine and gross motor function.

46. a. The body's pH control mechanism is so effective that blood pH stays within a narrow range: 7.35 to 7.45. Within the blood, values below 7.35 indicate acidity; values above 7.45 indicate alkalinity.

47. a. The interaction of oestrogen and progesterone with prolactin and oxytocin stimulates lactation. Corticotropin causes the foetus to secrete cortisol. FSH affects sexual development. Growth hormone influences growth and development.

48. c. Diffusion is the movement of solutes from an area of higher concentration to one of lower concentration. Osmosis is the movement of fluid from an area of lower solute concentration into an area of higher solute concentration. Active transport involves using energy to move a substance across a cell membrane. Endocytosis is an active transport method in which a cell engulfs a substance.

49. a. Glucagon is produced by pancreatic alpha cells and has the opposite effect of insulin. By raising blood glucose levels, it stimulates the liver to break down glycogen into glucose, elevating blood glucose levels. When blood glucose levels are low, a negative feedback mechanism stimulates alpha cells to release glucagon while high glucose levels reduce glucagon secretion. Pancreatic beta cells produce insulin which stimulates the liver to store glucose as glycogen removing glucose from the blood and reducing blood glucose levels. A negative feedback mechanism, sensitive to blood glucose concentrations, regulates secretion of insulin. The delta cells produce somatostatin which inhibits secretion of insulin and glucagon, as well as slowing down the absorption of nutrients from the gut. F cells secrete pancreatic polypeptide which inhibits the secretion of somatostatin and secretion of pancreatic digestive enzymes.

50. b. Creatinine is a break-down product of creatine phosphate in muscle and is usually produced at a constant rate that depends on muscle mass. Creatinine is chiefly filtered from the blood by the kidneys, although a small amount is actively secreted by the kidneys into the urine. There is little or no tubular reabsorption of creatinine. If the filtering capacity of the kidney is deficient, blood creatinine levels rise. Therefore, creatinine levels in blood and urine may be used to calculate creatinine clearance, which reflects the glomerular filtration rate (GFR) and renal function.

Glossary

abdomen: area of the body between the diaphragm and pelvis

abduct: to move away from the midline of the body; the opposite of *adduct*

acetabulum: hip joint socket into which the head of the femur fits

acromion: bony projection of the scapula

adduct: to move towards the midline of the body; the opposite of *abduct*

adenoids: paired lymphoid structures located in the nasopharynx; also known as the pharangeal tonsils

adrenals: a pair of glands one of which lies superior to each kidney; consists of a medulla and a cortex

afferent neurone: nerve cell that conveys impulses from the periphery to the central nervous system; the opposite of *efferent neurone*

alveolus: small saclike dilation of the terminal bronchioles of the lung

ampulla: saclike dilation of a tube or duct

anterior: front or ventral; the opposite of *posterior* or *dorsal*

antibody: immunoglobulin produced by B lymphocytes in response to exposure to a specific foreign substance (antigen)

antigen: foreign substance that causes antibody formation when introduced into the body

anus: distal end or outlet of the rectum

aorta: main trunk of the systemic arterial circulation, originating from the left ventricle and eventually branching into the two common iliac arteries

arachnoid: delicate middle membrane of the meninges

areola: pigmented ring around the nipple

arteriole: small branch of an artery

artery: vessel that carries blood away from the heart

arthrosis: joint or articulation

atrium: chamber or cavity

auricle: part of the external ear composed of cartilage and covered by skin that projects from the head at the external auditory meatus; also known as the pinna

axon: extension of a nerve cell that conveys impulses away from the cell body

bladder: any of a number of hollow organs that hold fluids or secretions

bone: dense, hard connective tissue that forms the skeleton

bone marrow: soft tissue in the cancellous bone of the epiphyses; crucial for blood cell formation and maturation

bronchiole: small branch of the bronchus

bronchus: larger air passage of the lung

buccal: pertaining to the cheek

bursa: fluid-filled sac lined with synovial membrane

caecum: pouch located at the proximal end of the large intestine

capillary: microscopic blood vessel that links arterioles with venules

carpal: pertaining to the wrist

cartilage: connective supporting tissue occurring mainly in the joints, thorax, larynx, trachea, nose and ear

central nervous system: one of the two main divisions of the nervous system; consists of the brain and spinal cord

cerebellum: portion of the brain situated in the posterior cranial fossa, behind the brain stem; coordinates voluntary muscular activity

cerebrum: largest and uppermost section of the brain, divided into hemispheres

cilia: small, hairlike projections on the outer surfaces of some cells

cochlea: spiral tube that makes up a portion of the inner ear

coeliac: pertaining to the abdomen

colon: part of the large intestine that extends from the caecum to the rectum

condyle: rounded projection at the end of a bone

contralateral: on the opposite side; the opposite of *ipsilateral*

cornea: convex, transparent anterior portion of the eye

coronary: pertaining to the heart or its arteries

cortex: outer part of an internal organ; the opposite of *medulla*

costal: pertaining to the ribs

cricoid: ring-shaped cartilage found in the larynx

cutaneous: pertaining to the skin

deltoid: shaped like a triangle (as in the deltoid muscle)

dendrite: branching process extending from the neuronal cell body that directs impulses towards the cell body

dermis: skin layer beneath the epidermis

diaphragm: membrane that separates one part from another; the muscular partition separating the thorax and abdomen

diaphysis: shaft of a long bone

diarthrosis: freely movable joint

diencephalon: part of the brain located between the cerebral hemispheres and the midbrain

distal: far from the point of origin or attachment; the opposite of *proximal*

diverticulum: outpouching from a tubular organ such as the intestine

dorsal: pertaining to the back or posterior; the opposite of *ventral* or *anterior*

duct: passage or canal

duodenum: shortest and widest portion of the small intestine, extending from the pylorus to the jejunum

dura mater: outermost layer of the meninges

ear: organ of hearing

efferent neurone: nerve cell that conveys impulses from the central nervous system to the periphery; the opposite of *afferent neurone*

endocardium: interior lining of the heart

endocrine: pertaining to secretion into the blood or lymph rather than into a duct; the opposite of *exocrine*

epidermis: outermost layer of the skin; lacking vessels

epiglottis: cartilaginous structure overhanging the larynx that guards against entry of food into the lungs

epiphyses: ends of a long bone

erythrocyte: red blood cell

exocrine: pertaining to secretion into a duct; the opposite of endocrine

eye: one of two organs of vision

fontanel: incompletely ossified area of a neonate's skull

foramen: opening

fossa: hollow or cavity

fundus: base of a hollow organ; the part farthest from the organ's outlet

gallbladder: excretory sac situated on the inferior surface of the liver's right lobe

ganglion: cluster of nerve cell bodies found outside the central nervous system

genitalia: reproductive organs of either the male or female, particularly the external parts of the reproductive system

gland: organ or structure in the body that secretes or excretes substances

glomerulus: compact cluster of capillaries in a kidney nephron

gonad: sex gland in which reproductive cells form (ovaries and testes)

haemoglobin: protein found in red blood cells that contains iron and transports oxygen

heart: muscular, cone-shaped organ that pumps blood throughout the body

hormone: substance secreted by an endocrine gland that triggers or regulates the activity of an organ or cell group

hyoid: small isolated U-shaped bone in the neck, which lies at the base of the tongue and to which it gives support

hypothalamus: structure in the diencephalon that secretes vasopressin (ADH) and oxytocin

ileum: distal part of the small intestine extending from the jejunum to the caecum

incus: one of three bones in the middle ear

inferior: lower; the opposite of *superior*

intestine: portion of the GI tract that extends from the stomach to the anus

intima: innermost structure

ipsilateral: on the same side; the opposite of *contralateral*

jejunum: one of three portions of the small intestine; connects proximally with the duodenum and distally with the ileum

joint: fibrous, cartilaginous, or synovial connection between bones

kidney: one of two urinary organs on the dorsal part of the abdomen

labia: lip-shaped structures; usually used to describe external female genitalia; part of the vulva

lacrimal: pertaining to tears

larynx: voice organ; joins the pharynx and trachea

lateral: pertaining to the side; the opposite of *medial*

leucocyte: white blood cell

ligament: band of white fibrous tissue that connects bones

liver: large gland in the right upper abdomen; divided into four lobes

lobe: defined portion of any organ, such as the liver or brain

lobule: small lobe

lumbar: pertaining to the area of the back between the thorax and the pelvis

lungs: organs of respiration found in the chest's lateral cavities

lymph: watery fluid in lymphatic vessels

lymph node: small oval structure that filters lymph, fights infection and aids haematopoiesis

lymphocyte: white blood cell; the body's immunologically competent cells

malleolus: projections at the distal ends of the tibia and fibula

malleus: tiny hammer-shaped bone in the middle ear

mammary: pertaining to the breast

manubrium: upper part of the sternum

meatus: opening or passageway

medial: pertaining to the middle; the opposite of *lateral*

mediastinum: middle portion of the thorax between the pleural sacs that contain the lungs

medulla: inner portion of an organ; the opposite of *cortex*

medulla oblongata: an extension of the upper part of the spinal cord contained within the skull and which holds the centres responsible for the regulation of the heart and blood vessels, respiration, salivation and swallowing

membrane: thin layer or sheet

metacarpals: bones of the hand located between the carpal bones and the fingers

metatarsals: bones of the foot located between the tarsal bones and the toes

midbrain: the small part of the brainstem that connects the hindbrain to the forebrain

muscle: fibrous structure whose contraction initiates movement

myocardium: thick, contractile layer of muscle cells that forms the heart wall

nares: nostrils

nephron: structural and functional unit of the kidney

nerve: bundle of fibres that convey impulses to and from the central nervous system and the body

neurone: nerve cell

neutrophil: white blood cell that removes and destroys bacteria, cellular debris and solid particles

occiput: back of the head

oesophagus: muscular canal that transports nutrients from the pharynx to the stomach

olfactory: pertaining to the sense of smell

ophthalmic: pertaining to the eye

ossicle: small bone, especially of the ear

ovary: one of two female reproductive organs found on each side of the lower abdomen, next to the uterus

palate: roof of the mouth

pancreas: secretory gland located in the left hypochondriac and epigastric regions

parotid: located near the ear (as in the parotid gland)

patella: bone that forms the kneecap

pectoral: pertaining to the chest or breast

pelvis: funnel-shaped structure; lower part of the trunk

pericardium: fibroserous sac that surrounds the heart and the origin of the great vessels

phalanx: one of the tapering bones that makes up the fingers and toes

pharynx: tubular passageway that extends from the base of the skull to the oesophagus

phrenic: pertaining to the diaphragm

pia mater: innermost covering of the brain and spinal cord

pituitary gland: gland attached to the hypothalamus that stores and secretes hormones

plantar: pertaining to the sole of the foot

plasma: fluid portion of blood

platelet: small, disc-shaped cell structure necessary for coagulation; also known as a thrombocyte

pleura: the covering of the lungs (visceral pleura) and of the inner surface of the chest wall (parietal pleura)

plexus: network of nerves, lymphatic vessels, or veins

pons: portion of the brain that lies between the medulla oblongata and the midbrain

popliteal: pertaining to the back of the knee

posterior: back or dorsal; the opposite of *anterior* or *ventral*

pronate: to turn the palm downwards; the opposite of *supinate*

prostate: male gland that surrounds the bladder neck and urethra

proximal: situated nearest the centre of the body; the opposite of *distal*

pupil: circular opening in the iris of the eye through which light passes

reflex: involuntary action

renal: pertaining to the kidney

scrotum: skin pouch that houses the testes and parts of the spermatic cords

semen: male reproductive fluid consisting of sperm cells and various supporting secretions

sphenoid: a bone forming the base of the cranium behind the eyes

spleen: highly vascular organ located in the left hypochondriac region

stapes: tiny stirrup-shaped bone in the middle ear

sternum: long, flat bone that forms the anterior portion of the thorax

stomach: major digestive organ, located in the epigastric, umbilical and left hypochondriac regions of the abdomen

striated: marked with parallel lines such as striated muscle

superior: higher; the opposite of *inferior*

supinate: to turn the palm of the hand upwards; the opposite of *pronate*

symphysis: growing together; a type of cartilaginous joint in which fibrocartilage firmly connects opposing surfaces

synapse: point of contact between adjacent neurones

systole: contraction of the heart muscle

talus: anklebone

tarsus: the seven bones of the ankle and proximal part of the foot

tendon: band of fibrous connective tissue that attaches a muscle to a bone

testis: one of two male gonads that produce semen

thalamus: one of two masses of grey matter that lie deep in the cerebral hemispheres in each side of the forebrain; relay points for all sensory messages that enter the brain

thyroid: secretory gland located at the front of the neck

tibia: shinbone

tongue: chief organ of taste, attached in the floor of the mouth

trachea: nearly cylindrical tube in the neck, extending from the larynx to the bronchi, that serves as a passageway for air

turbinate: shaped like a cone or spiral; a bone located in the posterior nasopharynx

ureter: either of a pair of tubes that transport urine from the kidney pelvis to the bladder

urethra: tubular structure that drains urine from the bladder

uterine tube: one of two ducts extending from the uterus to the ovary

uterus: internal female reproductive organ in which the fertilised ovum is implanted and the foetus develops

uvula: tissue projection that hangs from the soft palate

vagina: the canal in the female extending from the vulva to the cervix

valve: structure that permits fluid to flow in only one direction

vein: vessel that carries blood to the heart

vena cava: one of two large veins that returns blood from the peripheral circulation to the right atrium

ventral: pertaining to the front or anterior; the opposite of *dorsal* or *posterior*

ventricles: small cavities, such as those found in the brain or one of the two lower chambers of the heart

venule: small vessel that connects a vein and capillary plexuses

vertebra: any of the bones that make up the spinal column

viscera: internal organs

xiphoid: sword-shaped; the lower portion of the sternum

Study cards

The study cards in the following section are designed to test your knowledge of the structures of many of the organs, systems and mechanisms of action that occur in the body. Try to identify each of the indicated features shown in each diagram.

Answers are given at the bottom of the relevant page and further details are available with respect to each feature, within the main text.

Locating body cavities

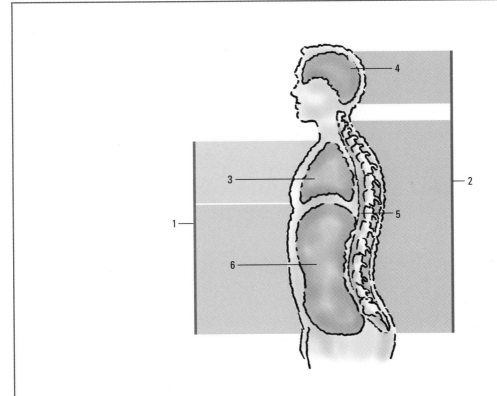

1. Ventral cavity   2. Dorsal cavity   3. Thoracic cavity   4. Cranial cavity   5. Vertebral cavity   6. Abdominopelvic cavity

Inside the cell

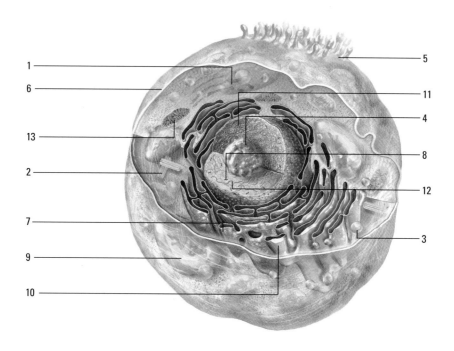

1. Cytoplasm   2. Centriole   3. Lysosome   4. Nucleolus   5. Microvilli   6. Cell membrane   7. Endoplasmic reticulum
8. Ribonucleic acid   9. Ribosomes   10. Golgi complex   11. Nucleus   12. Chromatin   13. Mitochondrion

Viewing the major skeletal muscles

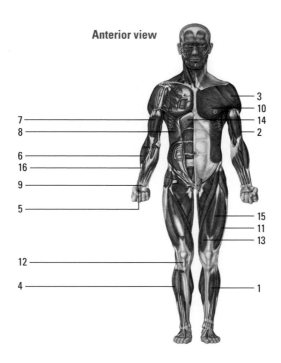

Anterior view

1. Tibialis anterior   2. Brachialis   3. Deltoid   4. Soleus   5. Flexor retinaculum   6. Flexor pollicis longus   7. Biceps brachii
8. External abdominal oblique   9. Pronator quadratus   10. Pectoralis major   11. Vastus lateralis   12. Patellar ligament
13. Vastus medialis   14. Rectus abdominis   15. Vastus intermedius   16. Abductor pollicis longus

Viewing the major skeletal muscles

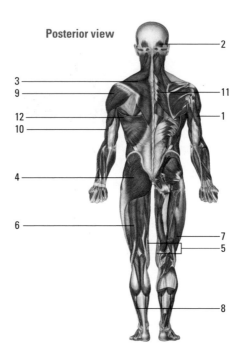

Posterior view

2

3
9

11

12
10

1

4

6

7
5

8

1. Triceps brachii   2. Occipitalis   3. Trapezius   4. Gluteus maximus   5. Sartorius   6. Biceps femoris   7. Vastus lateralis
8. Gastrocnemius   9. Deltoid   10. Brachialis   11. Rhomboid major   12. Latissimus dorsi

Muscle structure up close

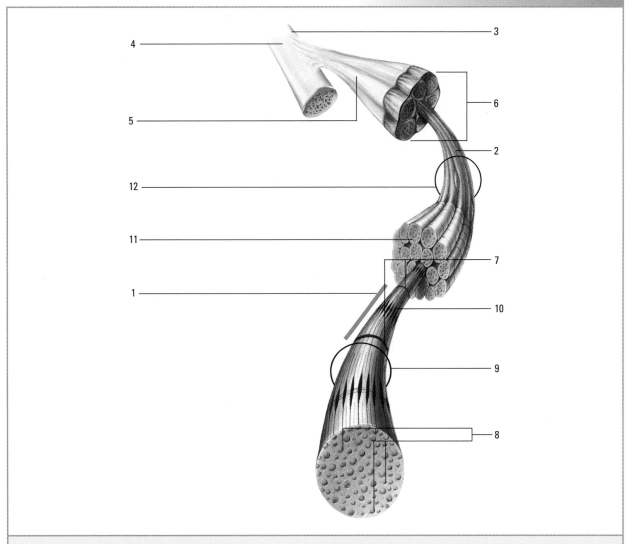

1. Sarcomere   2. Perimysium   3. Tendon   4. Periosteum   5. Epimysium   6. Belly   7. Z-line   8. Myosin and actin   9. Muscle fibre
10. H-zone   11. Endomysium   12. Fascicle

Viewing the major bones

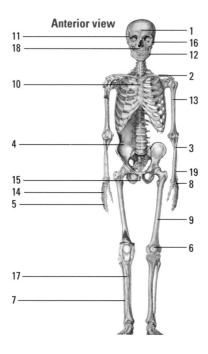

Anterior view

1. Frontal   2. Clavicle   3. Radius   4. Ilium   5. Phalanges   6. Patella   7. Fibula   8. Carpals   9. Femur   10. Sternum   11. Temporal
12. Mandible   13. Humerus   14. Metacarpals   15. Pubic symphysis   16. Zygomatic   17. Tibia   18. Maxilla   19. Ulna

Viewing the major bones

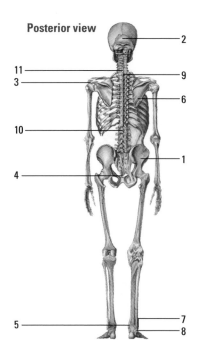

Posterior view

1. Ilium   2. Occipital   3. Acromion   4. Coccyx   5. Medial malleolus   6. Scapula   7. Lateral malleolus   8. Talus   9. T1
10. L1   11. C7

Parts of a neurone

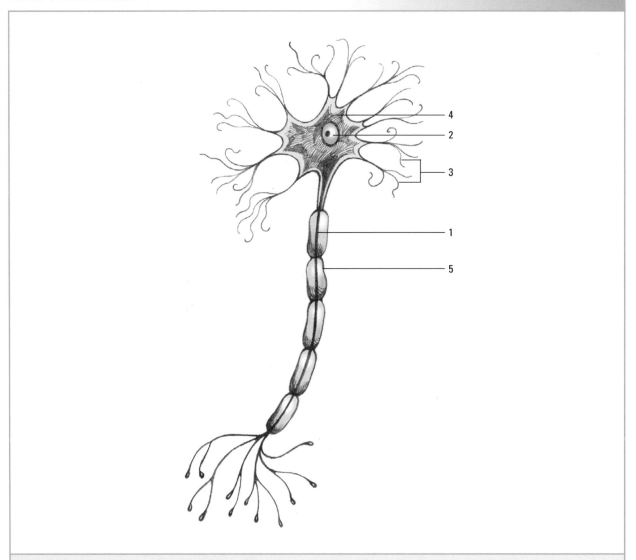

1. Axon   2. Nucleus of cell body   3. Dendrites   4. Cell body   5. Myelin sheath

The reflex arc

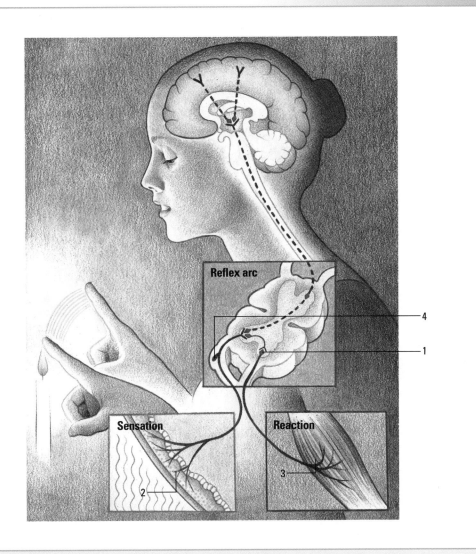

1. Interneurone 2. Sensory neurone 3. Motor neurone 4. Dorsal root ganglion

A close look at major brain structures

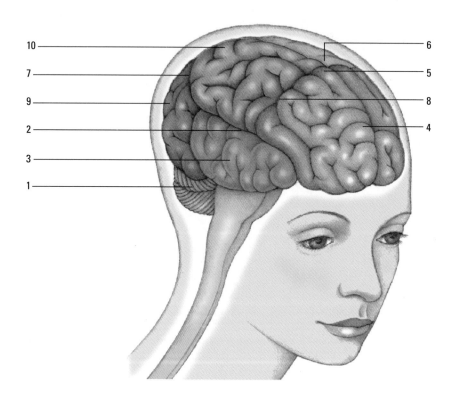

1. Cerebellum   2. Lateral sulcus   3. Temporal lobe   4. Frontal lobe   5. Motor cortex   6. Sensory cortex   7. Parieto-occipital fissure
8. Central sulcus   9. Occipital lobe   10. Parietal lobe

The limbic system

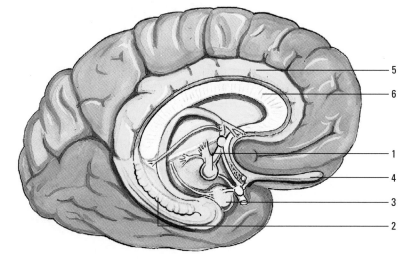

— 5
— 6
— 1
— 4
— 3
— 2

1. Mammillary body   2. Hippocampus   3. Amygdala   4. Olfactory tract   5. Cingulate gyrus   6. Corpus callosum

Arteries of the brain

Inferior view

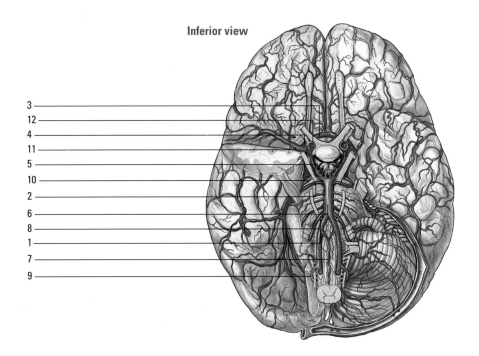

3
12
4
11
5
10
2
6
8
1
7
9

1. Vertebral artery   2. Superior cerebellar artery   3. Anterior communicating artery   4. Anterior cerebral artery   5. Posterior communicating artery   6. Basilar artery   7. Anterior spinal artery   8. Anterior inferior cerebellar artery   9. Posterior spinal artery   10. Posterior cerebral artery   11. Middle cerebral artery   12. Left internal carotid artery

A look inside the spinal cord

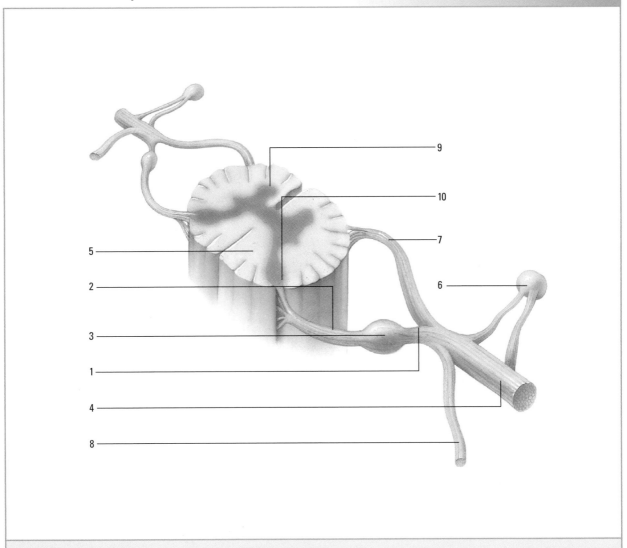

1. Spinal nerve   2. Posterior root   3. Posterior root (spinal) ganglion   4. Anterior ramus   5. Spinal cord   6. Sympathetic ganglion
7. Anterior root   8. Posterior ramus   9. Anterior horn   10. Posterior horn

The spinal nerves

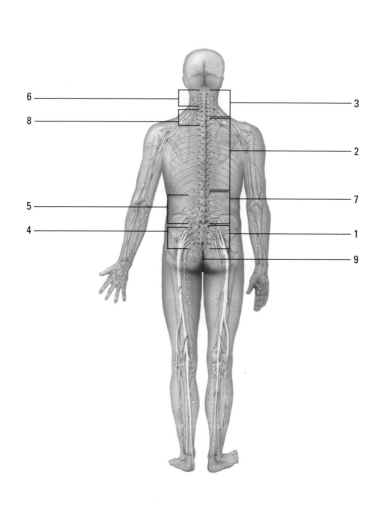

1. Sacral nerves (S1 to S5)   2. Thoracic nerves (T1 to T12)   3. Cervical nerves (C1 to C8)   4. Sacral plexus   5. Lumbar plexus
6. Cervical plexus   7. Lumbar nerves (L1 to L5)   8. Brachial plexus   9. Coccygeal nerve

Exit points for the cranial nerves

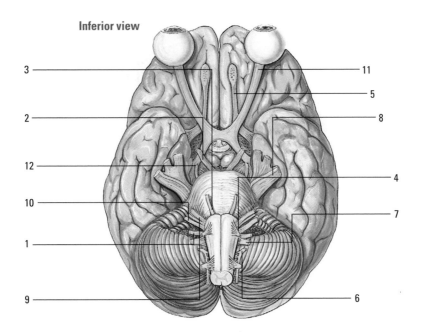

Inferior view

3
2
12
10
1
9

11
5
8
4
7
6

1. Glossopharyngeal (CN IX)   2. Oculomotor (CN III)   3. Abducens (CN VI)   4. Hypoglossal (CN XII)   5. Olfactory (CN I)   6. Spinal accessory (CN XI)   7. Vagus (CN X)   8. Trigeminal (CN V)   9. Facial (CN VII)   10. Vestibulocochlear (CN VIII)   11. Optic (CN II)   12. Trochlear (CN IV)

Intraocular structures

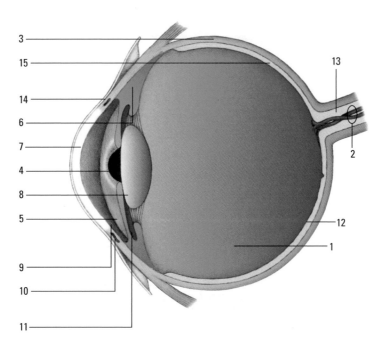

1. Vitreous humor   2. Central retinal artery and vein   3. Sclera   4. Pupil   5. Iris   6. Ciliary body   7. Cornea   8. Lens   9. Anterior chamber (filled with aqueous humour)   10. Schlemm's canal   11. Posterior chamber (filled with aqueous humuor)   12. Retina   13. Optic nerve   14. Conjunctiva (bulbar)   15. Choroid

Ear structures

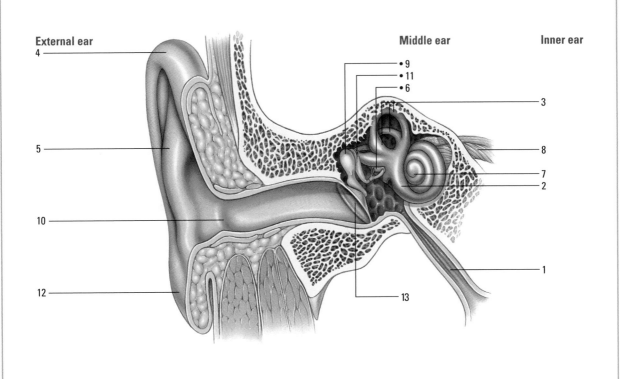

External ear   Middle ear   Inner ear

1. Eustachian tube   2. Vestibule   3. Semicircular canals   4. Helix   5. Anthelix   6. Stapes (stirrup)   7. Cochlea   8. Vestibulocochlear nerve   9. Malleus (hammer)   10. External acoustic meatus   11. Incus (anvil)   12. Lobule of auricle   13. Tympanic membrane (eardrum)

Components of the endocrine system

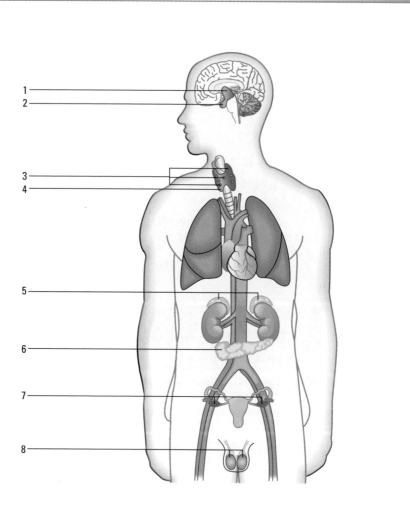

1. Hypothalamus   2. Pituitary   3. Parathyroids   4. Thyroid   5. Adrenals   6. Pancreas   7. Ovaries (female)   8. Testes (male)

Inside the heart

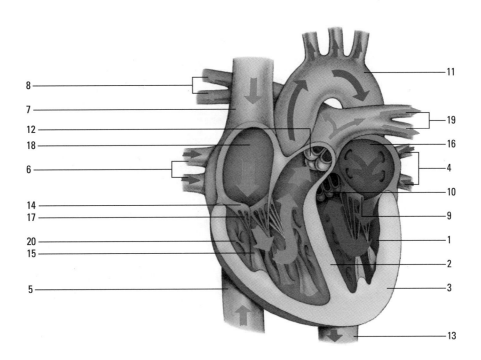

1. Left ventricle   2. Interventricular muscle   3. Myocardium   4. Left pulmonary veins   5. Inferior vena cava   6. Right pulmonary veins
7. Superior vena cava   8. Branches of right pulmonary artery   9. Mitral valve   10. Aortic semilunar valve   11. Aortic arch
12. Pulmonary semilunar valve   13. Descending aorta   14. Tricuspid valve   15. Papillary muscle   16. Left atrium   17. Chordae tendineae   18. Right atrium   19. Branches of left pulmonary artery   20. Right ventricle

Cardiac conduction system

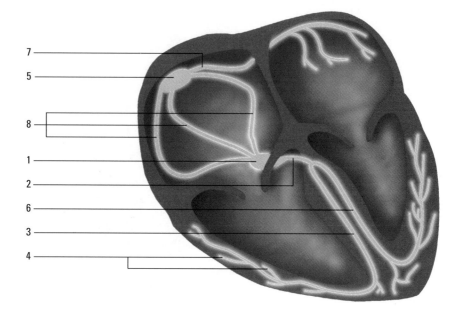

1. AV node   2. Bundle of His   3. Left bundle branch   4. Purkinje fibres   5. SA node   6. Right bundle branch   7. Interatrial bundle
8. Internodal tracts (posterior, middle and anterior)

Major blood vessels

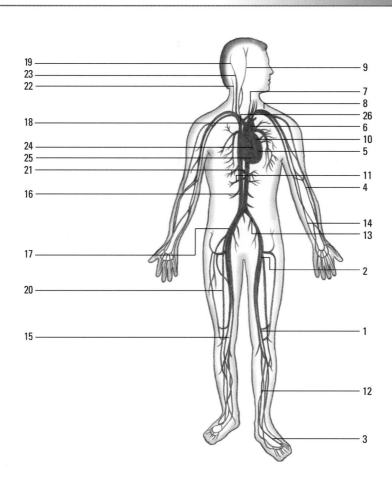

1. Popliteal artery   2. Femoral artery   3. Dorsalis pedis artery   4. Radial artery   5. Left ventricle   6. Aorta   7. Left common carotid artery   8. Left subclavian artery   9. Temporal artery   10. Left atrium   11. Renal arteries   12. Posterior tibial artery   13. Common iliac artery   14. Ulnar artery   15. Popliteal vein   16. Inferior vena cava   17. Common iliac vein   18. Superior vena cava   19. Transverse sinus   20. Femoral vein   21. Renal veins   22. Brachiocephalic vein   23. Right jugular vein   24. Right atrium   25. Right ventricle   26. Brachiocephalic artery

Vessels that supply the heart

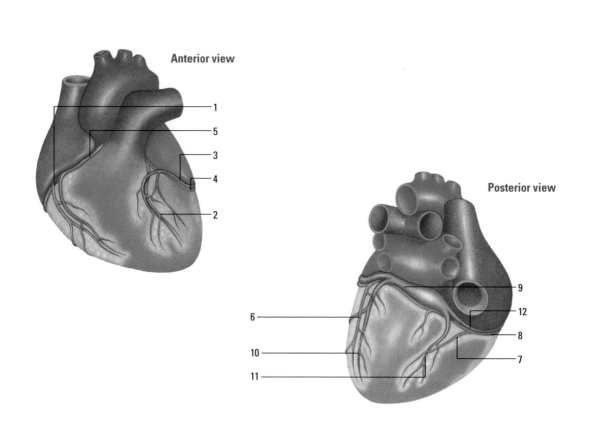

Anterior view

1
5
3
4
2

Posterior view

9
12
8
7
6
10
11

1. Small cardiac vein   2. Anterior interventricular (descending) branch of left main coronary artery   3. Great cardiac vein   4. Circumflex branch of left coronary artery   5. Right coronary artery   6. Circumflex branch of left coronary artery   7. Posterior interventricular (descending) branch of right coronary artery   8. Right coronary artery   9. Great cardiac vein   10. Posterior vein of left ventricle   11. Middle cardiac vein   12. Small cardiac vein

Organs and tissues of the immune system

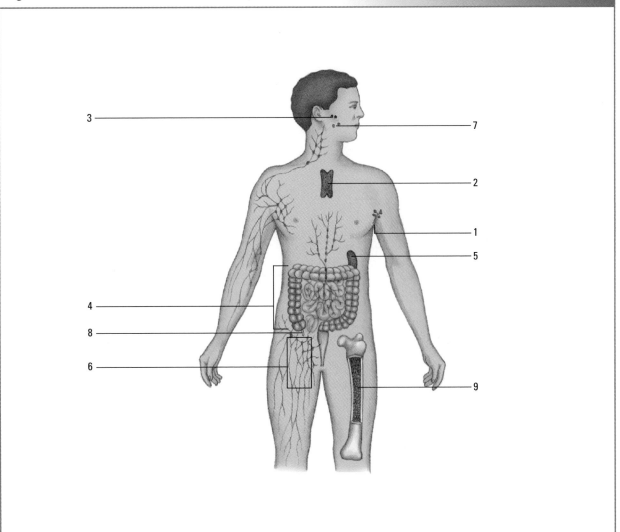

1. Lymph nodes 2. Thymus 3. Adenoids 4. Peyer's patches 5. Spleen 6. Lymphatic vessels and blood capillaries 7. Tonsils
8. Appendix 9. Bone marrow

Lymphatic vessels and lymph nodes

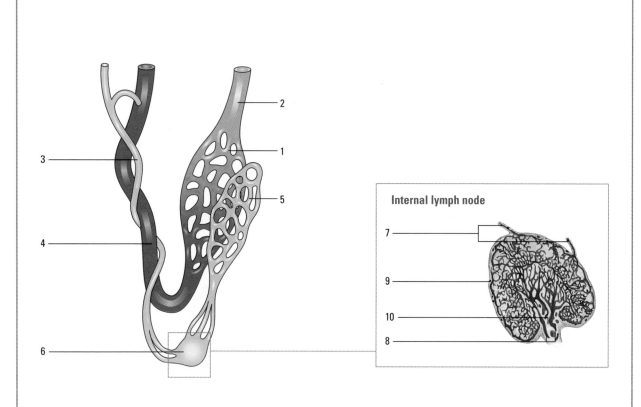

Internal lymph node

1. Blood capillaries   2. Vein   3. Lymphatic vessel   4. Artery   5. Lymphatic capillaries   6. Lymph node   7. Afferent vessels
8. Efferent vessels   9. Capsule   10. Hilum

Structures of the respiratory system

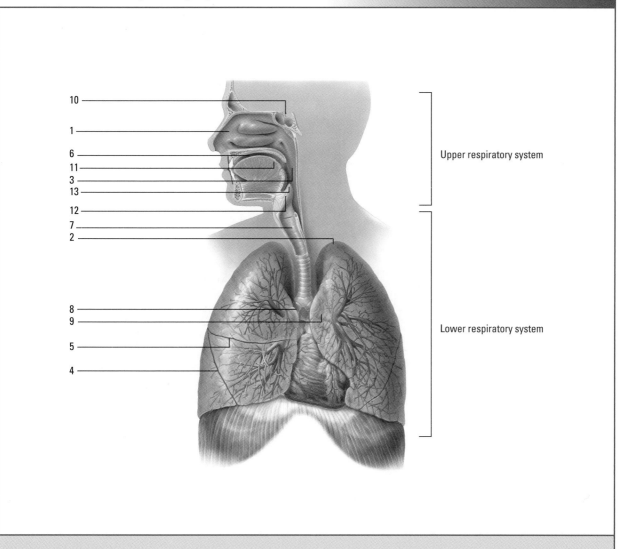

Upper respiratory system

Lower respiratory system

1. Nasal cavity   2. Apex of lung   3. Oropharynx   4. Oblique fissure   5. Horizontal fissure   6. Oral cavity   7. Trachea   8. Right main bronchus   9. Left main bronchus   10. Nasal sinus   11. Nasopharynx   12. Larynx   13. Laryngopharynx

A close look at a pulmonary airway

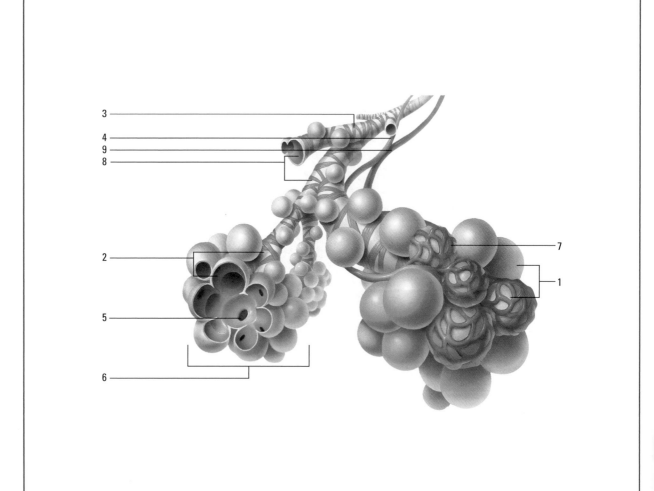

1. Alveoli   2. Alveolar duct   3. Smooth muscle   4. Pulmonary artery   5. Alveolar pore   6. Alveolar sac   7. Capillary beds
8. Respiratory bronchioles   9. Pulmonary vein

Structures of the GI system

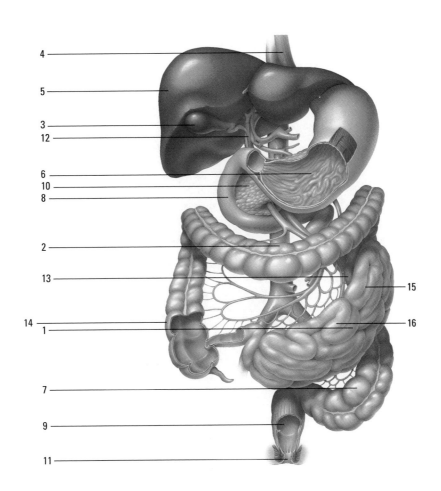

1. Ileum   2. Transverse colon   3. Gallbladder   4. Oesophagus   5. Liver   6. Stomach   7. Sigmoid colon   8. Duodenum   9. Rectum   10. Pancreas   11. Anus   12. Common bile duct   13. Descending colon   14. Ascending colon   15. Duodenum   16. Jejunum

Oral cavity

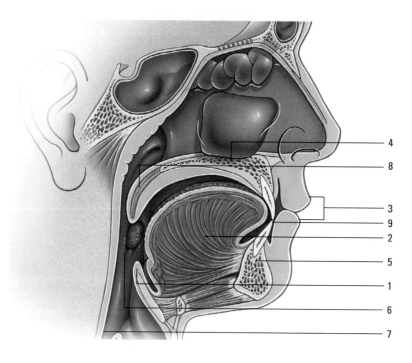

4
8

3
9
2

5
1
6
7

1. Epiglottis   2. Tongue   3. Lips   4. Hard palate   5. Mandible   6. Pharynx   7. Oesophagus   8. Soft palate   9. Teeth

A close look at the urinary system

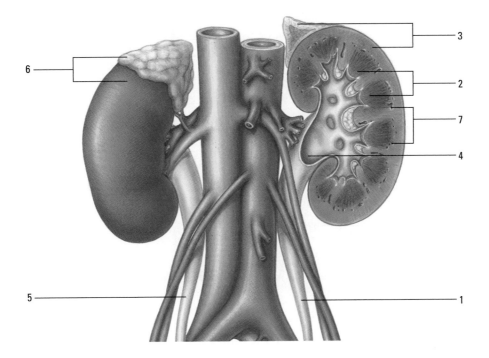

1. Left ureter   2. Renal papillae   3. Left kidney and adrenal gland (cross section)   4. Renal pelvis   5. Right ureter   6. Right kidney and adrenal gland   7. Renal parenchyma

Structure of the nephron

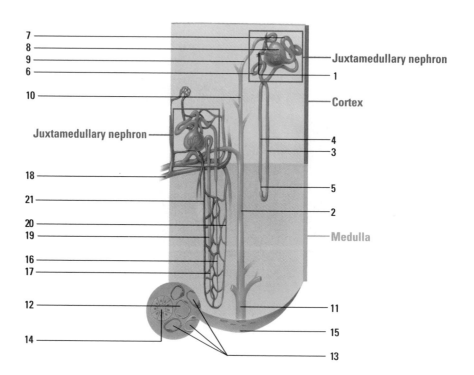

1. Macula densa   2. Medullary collecting duct   3. Distal straight tubule   4. Proximal straight tubule   5. Thin segment   6. Distal convoluted tubule   7. Proximal convoluted tubule   8. Renal corpuscle   9. Connecting tubule   10. Cortical collecting duct   11. Papillary duct   12. Descending thin limb of Loop of Henlé   13. Vasa recta   14. Distal straight tubule   15. Renal papilla   16. Descending thin limb of Loop of Henlé   17. Ascending thin limb of Loop of Henlé   18. Arcuate artery   19. Distal straight tubule   20. Venous rasa recta   21. Arterial vasa recta

A close look at the skin

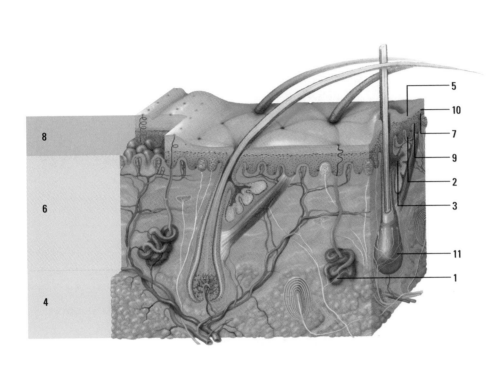

1. Eccrine sweat gland   2. Melanocyte   3. Sebaceous gland   4. Subcutaneous tissue   5. Stratum corneum   6. Dermis   7. Spinous layer (stratum spinosum)   8. Epidermis   9. Basal layer   10. Granular layer (stratum granulosum)   11. Hair bulb

A close look at the nail

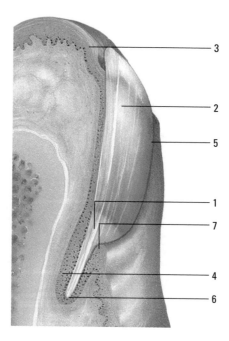

3

2

5

1

7

4

6

1. Lunula   2. Nail plate   3. Hyponychium   4. Nail root   5. Lateral nail fold   6. Nail matrix   7. Eponychium

Structures of the male reproductive system

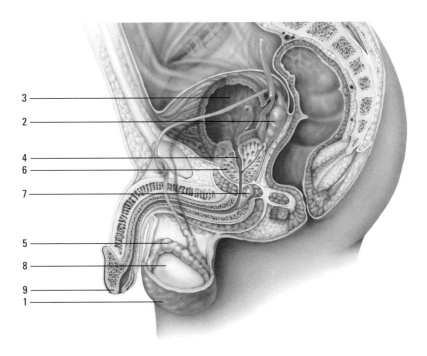

3
2
4
6
7
5
8
9
1

1. Scrotum   2. Seminal vesicles   3. Urinary bladder   4. Urethra   5. Epididymis   6. Prostate gland   7. Bulbourethral gland
8. Testis   9. Glans penis

Female external genitalia

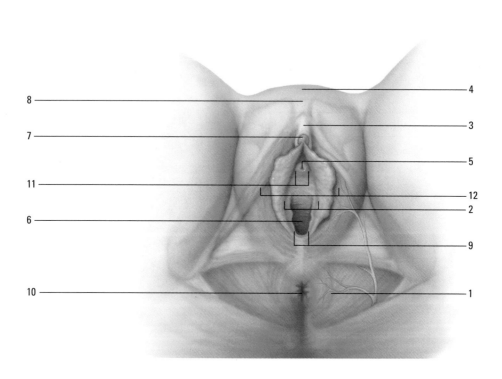

1. External anal sphincter   2. Labia minora   3. Prepuce   4. Mons pubis   5. Urethral orifice   6. Vaginal opening   7. Clitoris
8. Symphysis pubis   9. Openings of greater vestibular (Bartholin's) glands   10. Anus   11. Openings of paraurethral (Skene's) glands
12. Labia majora

Structures of the female reproductive system

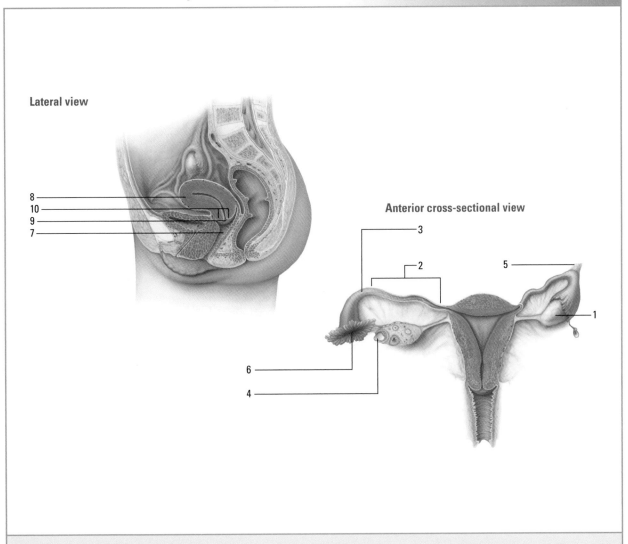

Lateral view

8
10
9
7

Anterior cross-sectional view

3
2
5
1
6
4

1. Ovary 2. Isthmus 3. Uterine tube 4. Secondary oocyte 5. Suspensory ligament of ovary 6. Abdominal opening of uterine tube 7. Vagina 8. Uterus 9. External cervical os 10. Cervix

The female breast

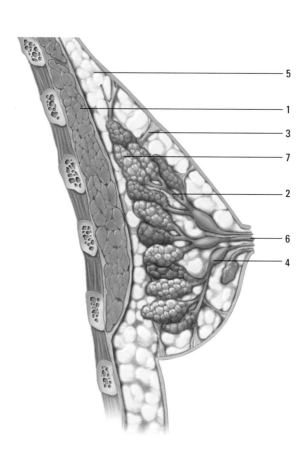

5

1

3

7

2

6

4

1. Pectoralis muscle   2. Lactiferous duct   3. Suspensory ligament   4. Ampulla   5. Fat   6. Nipple   7. Gland lobules

Selected references

Clemente, C. *Anatomy: Regional Atlas of the Human Body,* 6th ed. Philadelphia: Lippincott Williams & Wilkins, 2010.

Colbert, B., et al. *Anatomy and Physiology for Nursing and health Professionals,* 1st ed. New York: Pearson, 2009.

Drake, R., Vogl, A.W., and Mitchell, A.J.M. *Grays Anatomy for Students,* 2nd ed. New York: Churchill Livingstone, 2009.

Gould, R.E., and Dyer, R. *Pathophysiology for the Health Professions,* 4th ed. Philadelphia: W.B. Saunders Co., 2011.

Herlihy, B. *The Human Body in Health and Illness,* 4th ed. Philadelphia: W.B. Saunders Co., 2010.

Marieb, E.N., and Hoehn, K. *Human Anatomy and Physiology,* New York: Pearson, 2010.

Porth, C.M. *Essentials of Pathophysiology: Concepts of Altered Health State,* 8th ed. Philadelphia: Lippincott Williams & Wilkins, 2009.

Rogers, K.M., and Scott, W.N. *Nurses! Test Yourselves in Anatomy and Physiology,* London: McGraw-Hill, 2011.

Rohen, J.W., et al. *Color Atlas of Anatomy: A Photographic Study of the Human Body,* 7th ed. Philadelphia: Lippincott Williams & Wilkins, 2010.

Thibodeau, G., and Patton, K. *Anthony's Textbook of Anatomy and Physiology,* 19th ed. St. Louis: Mosby, 2009.

Tortora, G.J., and Derrickson, B.H. *Principles of Anatomy and Physiology,* 12th ed. New York: Wiley, 2008.

Waugh, A., and Grant, A. *Ross and Wilson Anatomy and Physiology in Health and Illness,* 11th ed. Edinburgh: Churchill Livingstone, 2010.

Index

Note: Illustrations are denoted by i.